Newnes Interfacing Companion

To Robert Winston Cheary,
friend and teacher.

Newnes
Interfacing Companion

A.C. Fischer-Cripps

Newnes

OXFORD AMSTERDAM BOSTON LONDON NEW YORK PARIS
SAN DIEGO SAN FRANCISCO SINGAPORE SYDNEY TOKYO

Newnes
An imprint of Elsevier Science
Linacre House, Jordan Hill, Oxford OX2 8DP
225 Wildwood Avenue, Woburn MA 01801-2041

First published 2002
Transferred to digital printing 2004
Copyright © 2002, A. C. Fischer-Cripps. All rights reserved

British Library Cataloguing in Publication Data
A catalogue record for this book is available from the British Library

Library of Congress Cataloguing in Publication Data
A catalogue record for this book is available from the Library of Congress

ISBN 0 750 65720 0

For information on all Newnes publications
visit our website at www.newnespress.com

Contents

Preface

Part 1: Transducers

Part 2: Interfacing

Preface

The overall aim of this book is to present transducer devices, computer interfacing and instrumentation electronics in a succinct and memorable fashion. The book combines physics, computer science and electrical engineering in a science/engineering context. Starting from the transfer of physical phenomena to electrical signals, the book presents a comprehensive treatment of computer interfacing and finishes with signal conditioning, data analysis and digital filtering. The book covers a wide scope but contains sufficient detail to allow a practical application of the theory. Detailed explanations are given, even of the most difficult of concepts. The review problems offer a level of complexity which provides sufficient challenge to impart a sense of achievement upon their completion. The accompanying project work reinforces the theoretical work while allowing the reader to gain the satisfaction and experience of actually constructing a working interfacing circuit that can be used on any personal computer with a serial port. The book will be useful for students who are new to the subject, and will serve as a handy reference for experienced engineers who wish to refresh their knowledge of a particular topic.

In writing this book, I was assisted and encouraged by many colleagues. In particular, I acknowledge the contributions of Alec Bendeli, Stephen Buck, Bob Graves, Walter Kalceff, Les Kirkup, Geoff Smith, Paul Walker, my colleagues at the University of Technology, Sydney, the staff of the CSIRO Division of Telecommunications and Industrial Physics, and all my former students. My sincere thanks to my wife and family for their unending encouragement and support. Finally, I thank Matthew Deans, Jodi Burton and the editorial and production teams at Newnes for their very professional and helpful approach to the whole publication process.

Tony Fischer-Cripps,
Killarney Heights, Australia, 2002

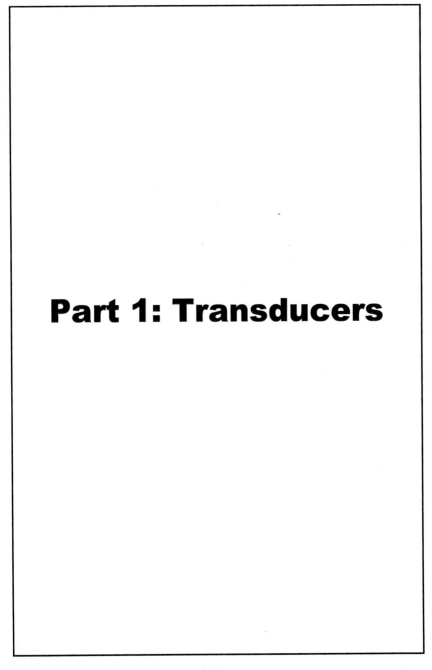

Part 1: Transducers

1.0 Transducers

A measurement system is concerned with the representation of one
physical phenomenon by another. The purpose of the measurement system
is for the measurement and control of a physical system.

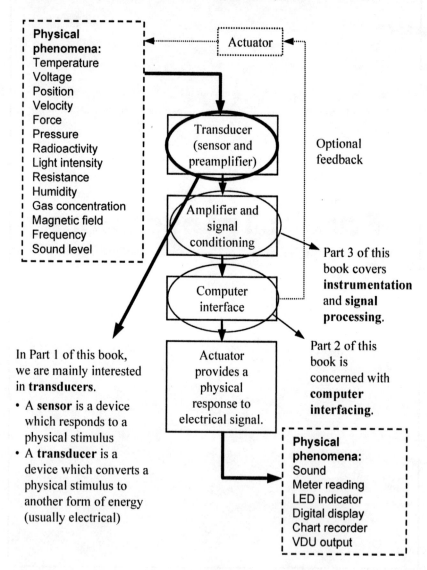

**Physical
phenomena:**
Temperature
Voltage
Position
Velocity
Force
Pressure
Radioactivity
Light intensity
Resistance
Humidity
Gas concentration
Magnetic field
Frequency
Sound level

Actuator

Transducer
(sensor and
preamplifier)

Optional
feedback

Amplifier and
signal
conditioning

Computer
interface

Part 3 of this
book covers
instrumentation
and **signal
processing**.

In Part 1 of this book,
we are mainly interested
in **transducers**.

- A **sensor** is a device
  which responds to a
  physical stimulus
- A **transducer** is a
  device which converts a
  physical stimulus to
  another form of energy
  (usually electrical)

Actuator
provides a
physical
response to
electrical signal.

Part 2 of this
book is
concerned with
**computer
interfacing**.

**Physical
phenomena:**
Sound
Meter reading
LED indicator
Digital display
Chart recorder
VDU output

1.1 Measurement systems

1.1.1 Transducers

Of most interest are the physical properties and performance
characteristics of a transducer. Some examples are given below:

Property	Method of measurement
Strain	Strain gauge, a resistive transducer whose resistance changes with length.
Temperature	Resistance thermometer, thermocouple, thermister, thermopile.
Humidity	Resistance change of hygroscopic material.
Pressure	Movement of the end of a coiled tube under pressure.
Voltage	Moving coil in a magnetic field.
Radioactivity	Electrical pulses resulting from ionisation of gas at low pressure.
Magnetic field	Deflection of a current carrying wire.

Performance characteristics

Static	Dynamic	Environmental
Sensitivity	Response time	Operating temperature
Zero offset	Damping	range
Linearity	Natural frequency	Orientation
Range	Frequency response	Vibration/shock
Span		
Resolution		
Threshold		
Hysteresis		
Repeatability		

A consideration of these characteristics influences the
choice of transducer for a particular application.
Further characteristics which are often important are
the operating life, storage life, power requirements
and safety aspects of the device as well as cost and
availability of service.

In industrial situations, the property being measured or controlled is called
the **controlled variable**. **Process control** is the procedure used to measure
the controlled variable and control it to within a tolerance level of a **set
point**. The controlled variable is one of several **process variables** and is
measured using a **transducer** and controlled using an **actuator**.

1.1.2 Methods of measurement

All measurements involve a comparison between a measured quantity and a reference standard. There are two fundamental methods of measurement:

Null method
- Direct comparison
- No loading
- Can be relatively slow

Deflection method
- Indirect comparison
- Deflection from zero until some balance condition achieved
- Limited in precision and accuracy
- Loading (transducer itself takes some energy from the system being measured)
- Relatively fast

Null method: Bridge circuit.

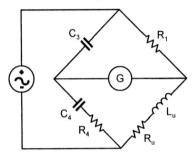

An unknown component is inserted into the bridge and the values of the others are altered to achieve balance condition. At balance, no current flows through the galvanometer G.

$$R_1 R_4 = \frac{L_u}{C_3}$$

$$\frac{R_1}{C_4} = \frac{R_u}{C_3}$$

Deflection method: Moving coil voltmeter.

magnet

pointer

coil

Although such a meter is designed to have a very high internal impedance, it has to draw some current from the circuit being measured in order to cause a deflection of the pointer. This may affect the operation of the circuit itself and lead to inaccurate readings – especially if the output resistance of the voltage source being measured is large.

1.1.3 Sensitivity

An important parameter associated with every transducer is its **sensitivity**.
This is a measure of the magnitude of the output divided by the magnitude
of the input.

$$\text{sensitivity} = \frac{\text{output signal}}{\text{input signal}}$$

$$= \frac{dO}{dI}$$

> e.g. The sensitivity of a thermocouple may
> be specified as 10 μV/^{o}C indicating that for
> each degree change in temperature between
> the sensor and the "reference" temperature,
> the output signal changes by 10 μV. The
> sensitivity may not be a constant across the
> working range.

The output voltage of most transducers is in the millivolt range for
interfacing in a laboratory or light industrial applications. For heavy
industrial applications, the output is usually given as a current rather than a
voltage. Such devices are usually referred to as **"transmitters"** rather than
transducers.

In most applications, the chances are that the signal produced by the
transducer contains **noise**, or unwanted information. The proportion of
wanted to unwanted signal is called the **signal-to-noise ratio** or SNR
(usually expressed in decibels).

The higher the **SNR** the better. In
electronic apparatus, noise signals often
arise due to thermal random motion of
electrons and is called white noise.
White noise appears at all frequencies.

$$\text{SNR} = 20\log_{10}\left|\frac{V_S}{V_n}\right|$$

Signal voltage

Noise voltage

> The first stage of any amplification of signal
> is the most critical when dealing with noise.
> In most sensitive equipment, a **preamplifier**
> is connected very close to the transducer to
> minimise noise and the resulting amplified
> signal passed to a main, or power amplifier.

The noise produced by a transducer limits its ability to detect very small
signals. A measure of performance is the **detectivity** given by:

$$d = \frac{1}{\text{least detectable input}}$$

> e.g. If d = 10^6 V^{-1} for a voltmeter, it means
> that the device can measure a voltage as low
> as 10^{-6} V.

The least detectable input is often referred to as the **noise floor** of the
instrument. The magnitude of the noise floor may be limited by the
transducer itself or the effect of the operating environment.

1.1.4 Zero, linearity and span

The **range** of a transducer is specified by the maximum and minimum input and output signals.

> e.g. A thermocouple has an input range of –100 to +300 ºC and an output range of –1 to +10 mV.

The **span** or **full scale deflection (fsd)** is the maximum variation in the input or output:

$$S = O_{max} - O_{min}$$

> e.g. The thermocouple above has an input span of 400 ºC and an output span S of 11 mV.

span ← Output maximum and minimum

Zero and span calibration controls:

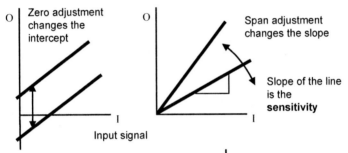

Zero adjustment changes the intercept

Span adjustment changes the slope

Slope of the line is the **sensitivity**

O Input signal I

The % of **non-linearity** describes the deviation of a linear relationship between the input and the output.

$$\text{Max non-linearity} = \frac{\delta}{S} \times 100$$

A linear output can be obtained by using a look-up table or altering the output signal electronically.

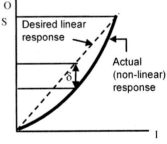

Desired linear response

Actual (non-linear) response

Zero offset errors can occur because of calibration errors, changes or ageing of the sensor, a change in environmental conditions, etc. The error is a constant over the range of the instrument.

A change in sensitivity, or a **span error**, results in the output being different to the correct value by a constant %. That is, the error is proportional to the magnitude of the output signal (change in slope).

1.1.5 Resolution, hysteresis and error

A continuous increase in the input signal sometimes results in a series of discrete steps in the output signal due to the nature of the transducer.

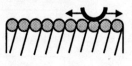

e.g. A wire wound potentiometer being used as a distance transducer. The wiper moves over the windings bringing a step change in resistance (R of one turn) with a change in distance.

The **resolution** of a transducer is defined as the size of the step divided by the **fsd** or span and is given in %.

$$\text{Resolution} = \frac{\delta O}{S}$$

e.g. The resolution of a 100 turn potentiometer is 1/100 = 1%.

For a particular input signal, the magnitude of the output signal may depend on whether the input is increasing or decreasing – this is called **hysteresis**.

Maximum hysteresis $= \dfrac{\delta}{S} \times 100$

In mechanical systems, hysteresis usually occurs due to backlash in moving parts (e.g. gear teeth).

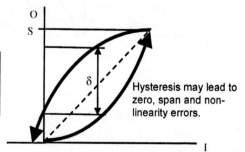

Hysteresis may lead to zero, span and non-linearity errors.

The general response of a transducer is usually given as a percent **error**.

$$\text{Error} = \frac{\delta}{S} \times 100$$

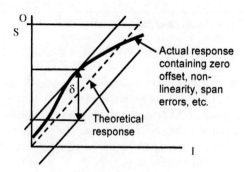

Actual response containing zero offset, non-linearity, span errors, etc.

Theoretical response

1.1.6 Fourier analysis

Analog input signals that require sampling by a digital to analog converter system do not usually consist of just a single sinusoidal waveform. Real signals usually have a variety of amplitudes and frequencies that vary with time.

Such signals can be broken down into component frequencies and amplitudes using a method called **Fourier analysis**. Fourier analysis relies on the fact that any periodic waveform, no matter how complicated, can be constructed by the superposition of sine waves of the appropriate frequency and amplitude.

For example, a square wave can be represented using the sum of individual component sine waves:

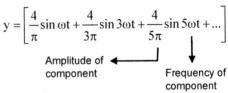

$$y = \left[\frac{4}{\pi} \sin \omega t + \frac{4}{3\pi} \sin 3\omega t + \frac{4}{5\pi} \sin 5\omega t + \dots \right]$$

Amplitude of component ←───┘

Frequency of component

Fourier analysis, or the breaking up of a signal into its component frequencies, is important when we consider the process of filtering and the conversion of an analog signal into a digital form.

$$y = \frac{4}{\pi} \sin \omega t$$

$$y = \left[\frac{4}{\pi} \sin \omega t + \frac{4}{3\pi} \sin 3\omega t \right]$$

$$y = \left[\frac{4}{\pi} \sin \omega t + \frac{4}{3\pi} \sin 3\omega t + \frac{4}{5\pi} \sin 5\omega t \right]$$

1.1.7 Dynamic response

The **dynamic response** of a transducer is concerned with the ability for the output to respond to changes at the input. The most severe test of dynamic response is to introduce a **step** signal at the input and measure the time response of the output.

Of particular interest are the following quantities:

- **Rise time**
- **Response time**
- **Time constant** τ

A step signal at the input causes the transducer to respond to an infinite number of component frequencies. When the input varies in a sinusoidal manner, the amplitude of the output signal may vary depending upon the frequency of the input if the frequency of the input is close to the **resonant frequency** of the system. If the input frequency is higher than the resonant frequency, then the transducer cannot keep up with the rapidly changing input signal and the output response decreases as a result.

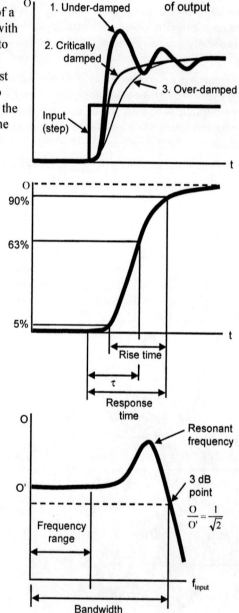

1.1.8 PID control

In many systems, a **servo feedback loop** is used to control a desired quantity. For example, a thermostat can be used in conjunction with an electric heater element to control the temperature in an oven. Such a servo loop consists of a sensor whose output controls the input signal to an actuator.

The difference between the target or **set point** and the current value of the controlled variable is the **error** signal Δe. If the error is larger than some preset tolerance or **error band**, then a **correction signal**, positive or negative, is sent to the actuator to cause the error to be reduced. In sophisticated systems, the error signal is processed by a **PID** controller before a correction signal is sent to the actuator. The PID controller determines the magnitude and type of the correction signal to be sent to the actuator to reduce the error signal.

The characteristics of a PID controller are expressed in terms of gains. The correction signal O from the PID controller to the actuator is given by the sum of the error Δe term multiplied by the proportional gain K_p, the integral gain, K_i and the derivative gain K_d.

$$O(t) = K_p \Delta e + K_I \int \Delta e \, dt + K_d \frac{d\Delta e}{dt}$$

- The **proportional** term causes the controller to generate a signal to the actuator whose amplitude is proportional to the magnitude of the error. That is, a large correction is made to correct a large error.
- The **integral** term is used to ramp the actuator to the final state to overcome friction or hysteresis in the system. It is a long-term correction and allows the system to servo to the target value.
- The **derivative** signal offers a damping response that reduces oscillation. The magnitude of the derivative correction depends upon the rate of change of the magnitude of the error signal. If the signal changes rapidly, a large correction is made.

The PID correction acts upon the error signal which is itself a function of time. The PID correction is thus also a function of time. For example, in servo **motion control**, a PID controller is able to cause the moving body (e.g. a robot arm) to accelerate, maintain a constant velocity, and decelerate to the target position.

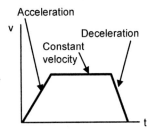

1.1.9 Accuracy and repeatability

Accuracy is a quantitative statement about the closeness of a measured
value with the **true value**.

The true value of a quantity is that
which is specified by international
agreement.

The kilogram is the unit of
mass and is equal to the
mass of an international
standard kilogram held in
Paris.

There is a difference between the
accuracy and the **precision** of physical
measurements.

High precision need not be
accompanied by high accuracy.
Precision is measured by the standard
deviation of several measurements.

High accuracy may also be
accompanied by a wide **scatter** in
the measurement readings leading
to low precision.

+ + + + + ● true value	High precision Low accuracy This condition could be caused by a **systematic error** in the measuring system (e.g. zero offset).
+ + + ● true value + + +	Low precision High accuracy This condition could be caused by a **random error** in the measuring system.
+ + + + ● + + + true value	High precision High accuracy

1.1.10 Mechanical models

The response of materials and systems can often be modelled by **springs** and **dashpots**. This allows both static and dynamic processes to be modelled mathematically with some convenience. Most materials have a mechanical character that falls somewhere in between the two extremes of a solid and a fluid. Springs represent the solid-like characteristics of a system. Dashpots represent the fluid-like aspects of a system.

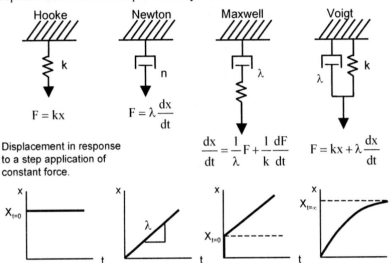

Hooke

$F = kx$

Displacement in response to a step application of constant force.

Newton

$F = \lambda \dfrac{dx}{dt}$

Maxwell

$\dfrac{dx}{dt} = \dfrac{1}{\lambda}F + \dfrac{1}{k}\dfrac{dF}{dt}$

Voigt

$F = kx + \lambda \dfrac{dx}{dt}$

Deflection of springs

If two or more springs are connected in parallel, then they experience a common displacement. In this case, the overall stiffness is given by:

$$k = \sum_{i=1}^{n} k_i$$

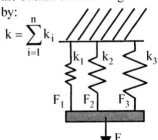

If two (or more) springs are connected in series, then loaded with a common force, then the total overall stiffness is given by:

$$k = \dfrac{1}{\displaystyle\sum_{i=1}^{n} \dfrac{1}{k_i}}$$

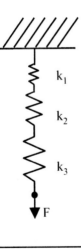

1.1.11 Review questions

1. A moving coil galvanometer has a series resistance of $R_M = 120 \ \Omega$ and a full-scale deflection at 2.5 µA. The display scale is divided into 100 equal divisions. The meter is to be used as a voltmeter to measure the emf from a 3.0 V source which has an output resistance R_O of 1.5 kΩ.

   (a) Determine the resistance R_S required to be connected in series with the galvanometer to make a voltmeter 0–5 V range.

   (b) Determine the uncertainty in the measured value for the above source emf due to meter resolution.

   (c) Determine the reading on the voltmeter when it is connected across the 3.0 V voltage source.   (Ans: 2 MΩ, 0.005 V, 2.9985 V)

2. An infrared gas analyser is used to measure the concentration of carbon monoxide (CO) in the exhaust gases of a motor vehicle. Before the measurement is taken, purified air containing no CO is introduced and the "zero" is adjusted for 0 mV on the output display. Then, a calibrated mixture of CO and air at 400 ppm is introduced and the span adjusted to give 400 mV on the output. The exhaust gas is then sampled by the instrument and the reading is 350 mV. It is discovered later that the concentration of the calibrated mixture was in error and should have been 410 ppm. Assuming the response of the instrument is linear, determine a corrected value for the measured concentration.   (Ans: 358.8)

3. The diagram shows the output of a linear transducer as a function of its input.

   (a) What formal term is given to the slope of this line?

   (b) What control is used to adjust this slope during calibration, the "zero" or the "span"?

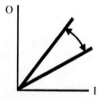

4. List three static, three dynamic and three environmental performance characteristics which would influence the choice of a transducer for a particular application.

5. A dashpot and spring in parallel (the Voigt model) can be represented by a resistor (k becomes R) and an inductor (λ becomes the inductance L) in series. If the applied force is replaced by voltage, and the resulting displacement is replaced by the current, show that the magnitude of the applied force and the magnitude of the displacement are related by:

$$\left|F\right| = \sqrt{k^2 + \omega^2 \lambda^2} \left|x\right|$$

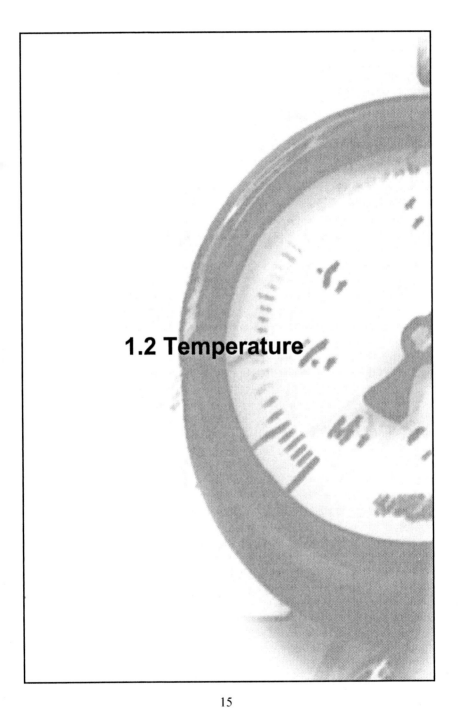

1.2 Temperature

1.2.1 Temperature

The measurement of temperature is naturally associated with the definition of a temperature scale.

Celsius temperature scale:
defined such that
0 °C = ice point of water
100 °C = boiling point of water.

Fahrenheit temperature scale:
defined such that
32 °F = ice point of water
212 °F = boiling point of water

Note: Standard atmospheric pressure (1 atm) is defined as 760 mm Hg (ρ = 13.5951 g/cm^3) at g = 9.80665 msec^{-2}.

$$ ^oC = \left({^oF} - 32 \right) \frac{100}{180} $$

The **International Temperature Scale** is based on the definition of a number of basic **fixed points**. The basic fixed points cover the range of temperatures to be normally found in industrial processes. They are (expressed here in °C):

1. Temperature of equilibrium between liquid and gaseous oxygen at 1 atm pressure is: −182.97 °C

2. Temperature of equilibrium between ice and air-saturated water at normal atmospheric pressure (ice point) is: 0.000 °C

3. Temperature of equilibrium between liquid water and its vapour at a pressure of 1 atm pressure (steam point) is: 100.000 °C

   Most common

4. Temperature of equilibrium between liquid sulphur and its vapour at 1 atm pressure is: 444.60 °C

5. Temperature of equilibrium between solid silver and liquid silver at normal atmospheric pressure is: 960.5 °C

6. Temperature of equilibrium between solid gold and liquid gold at normal atmospheric pressure is: 1063 °C

Other fixed points have been defined which facilitate calibration of thermometers in particular applications. Some examples are:

- Equilibrium between solid and gaseous CO_2: −78.5 °C
- Freezing mercury: −38.87 °C
- Freezing tin: 231.8 °C
- Freezing lead: 327.3 °C
- Freezing tungsten: 3400 °C

1.2.2 Standard thermometers

Temperatures in between the standard fixed points are found using standard thermometers which have been calibrated using the fixed points as follows:

From **−190 °C to the ice point**, the temperature is found from the resistance of a platinum resistance thermometer:

$$R_T = R_0 \left(1 + AT + BT^2 + C(T-100)^3\right)$$

The constants R_0, A, B and C, and degree of non-linearity are determined from the ice, steam, sulphur points and oxygen points.

From the **ice point to 660 °C**, the temperature is found from the resistance of a Platinum resistance thermometer:

$$R_T = R_0 \left(1 + AT + BT^2\right)$$

The constants R_0, A and B, and degree of non-linearity, are determined from the ice, steam and sulphur points.

From **600 °C to the gold point** (1063 °C) temperatures are found from the emf generated using a platinum/platinum-rhodium thermocouple where the cold junction is held at 0 °C. The temperature is found from:

$$emf = A + BT + CT^2$$

The constants A, B and C are determined from the freezing point of antimony, the silver and gold points.

Above the gold point, temperature is determined using a radiation pyrometer which compares the intensity of the light of a particular wavelength to that which would be emitted by a black body at temperature T.

Note: The official SI unit of temperature is the **Kelvin**. It is the temperature equal to the fraction 1/273.16 of the temperature of the triple point of water.

The triple point of water is the state of pure water existing as an equilibrium mixture of ice, liquid and vapour. Let the temperature of water at its triple point be equal to 273.16 K. This assignment corresponds to an ice point of 273.15 K or 0 °C − slightly lower than the triple point. The triple point is used as the standard fixed point because it is reproducible.

BS1041: 1943.

1.2.3 Industrial thermometers

In practice, thermometers used in industry have to be robust, reliable and often fast-acting. There are two general classes of thermometer, those that make contact with the body whose temperature to be measured, and those that do not.

Contact

- Expansion of solids (bimetallic strip) ——————→
- Expansion of liquids (mercury in glass)
- Expansion of gases (bellows)
- Thermoelectric junctions (thermocouple)
- Electrical resistance (thermistor)
- Change of state (melting point methods)

Bimetallic strip type thermometer

Non-contact

- Optical pyrometers (change in colour of hot bodies, disappearing filament device)
- Total radiation pyrometer (intensity of all wavelengths of radiation from hot body measured by focussing rays, using a lens or mirror, onto a "**receiver**" which may be a thermocouple or resistance element).

Thermocouple tip with
wires bonded together

The choice of thermometer depends on:

- The range of temperatures to be measured.
- Permissible time lag.
- Risk of chemical reaction with thermometer.
- Size and space requirements – ease of readings.
- Robustness.
- Single readings or recordings.

Precautions:

- Good contact between the hot body and sensor.
- Sensor to have small heat capacity.
- Chemical reactions which absorb or liberate heat to be avoided.
- Condensation to be avoided (latent heat may cause errors in temperature measurement).
- Electrical shielding to reduce noise pickup.

Bellows type
thermometer

1.2.4 Platinum resistance thermometer

The electrical resistance of a platinum wire-wound resistor changes with temperature. The response is reasonably linear and can be approximated by:

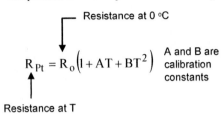

$$R_{Pt} = R_0\left(1 + AT + BT^2\right)$$

Resistance at 0 °C

A and B are calibration constants

Resistance at T

The resistance of the sensor changes with temperature. When the resistance changes, the current in the circuit changes. The rheostat is adjusted to bring the current back to its former value. This can be achieved by keeping the voltage across the standard resistor a constant using the rheostat. The change in voltage on the measuring potentiometer is thus due to a change in temperature only.

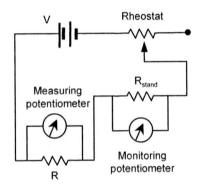

The Pt resistance sensor normally contains a supplementary ballast resistor (having a negligible change of resistance with temperature) the value of which is selected to make the total resistance of the element R_0 to be 100.0 Ω at 0 °C.

The change in resistance over a temperature range of 0 to 100 °C is called the **fundamental interval** and fixes the sensitivity of the device. A fundamental interval of 38.5 Ω is specified in BS1904 for temperature ranges up to 600 °C. Above 600 °C, the fundamental interval may be reduced to 10.000 Ω or even 1.000 Ω.

Power dissipated in element not to exceed 0.1 MΩ to avoid **self-heating**.

For a fundamental interval of 38.5 Ω, and $R_0 = 100$ Ω, the calibration constants for Pt are:

$$A = 3.91 \times 10^{-3} \, {}^{\circ}C^{-1}$$
$$B = -5.85 \times 10^{-7} \, {}^{\circ}C^{-2}$$

Pt resistance thermometer probe

1.2.5 Liquid-in-glass thermometer

A common thermometer in industry is the liquid-in-glass type which might contain either mercury or alcohol.

Advantages:
• Cheap, simple and portable.

Disadvantages:
• Restrictions on orientation
• High heat capacity.
• Significant time lag.

Type "A" thermometers are mercury-in-glass inert gas, solid stem. Type "B" are alcohol-in-glass, solid stem.

Designed for temperature range −120 °C to +510 °C and may be either total immersion or 100 mm immersion.

There are specific constructional guidelines (BS1704) which ensure uniformity of performance of thermometers from different manufacturers.

Constructional features:
• Stem: made of lead glass with an enamel back.
• Bulb: made cylindrical and has an external diameter not exceeding that of the stem.
• Thermometer is required to be annealed before graduation.
• Graduation lines are of uniform thickness not exceeding 0.15 mm and a line in a plane at right angles to the stem aligned to the left when the stem is viewed from the front in a vertical position.
• Immersion line is etched on the back of the stem for 100 mm immersion thermometers.
• A glass ring or rounded top is required at the top of the stem.
• A safety volume exists at the top of the capillary tube which is at least 20 mm above the top graduation line.

Safety gap

Markings:
• Gas filling employed, e.g. N_2.
• Manufacturer's mark.
• Schedule mark.

e.g. GP 150C/Total means general purpose thermometer, maximum temperature 150 °C, total immersion type.

Liquid-in-glass thermometer

1.2.6 Radiation pyrometer

Radiation **pyrometers** are usually used to measure high temperatures where physical contact with the hot body is not possible. A very popular form of pyrometer is the **disappearing filament** type.

The brightness of an electric filament lamp is adjusted by the operator by altering the current that passes through it. The hot body and an electric filament are both visible through an eyepiece. When brightness of the filament matches that of the hot body, the filament becomes invisible. The current through the filament at the **matching point** is an indication of temperature of the hot body.

Optical pyrometer

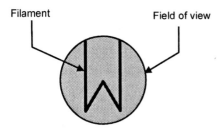

Filament   Field of view

Note: An additional screen may also be employed before the objective lens of the instrument to reduce the amount of incoming radiation. This permits a lower current to be used when measuring the temperature of very hot bodies and thus increasing filament life. With the screen in place (usually a piece of optically neutral glass) a second scale of temperatures is provided.

Usually, a **red filter** is used at the eyepiece so that matching is done at a particular wavelength (makes it easier to obtain a match). A correction table is used to obtain a true temperature from the indicated value which accounts for non-black body radiation when using the red filter.

Note: It is very common to use the eye as an optical pyrometer. For example, in the heat treatment of metals, it is sometimes required to heat until "cherry red" etc.

1.2.7 Thermocouple

A thermocouple consists of two **dissimilar metals** joined at either end. One of the junctions is held at a reference temperature, and the other junction is at the temperature to be measured. If a voltmeter is introduced into the circuit, the voltage depends on the difference in temperature between the two junctions of the device.

Sensor ("hot junction")

Hot junction

Reference ("cold junction")

Advantages:
- Able to measure high temperatures.
- Easily calibrated.
- Mechanically robust.
- Reasonably resistant to chemical attack.
- Can measure temperature of solids, liquids and gases.
- Reasonably fast response time (usually a few seconds).

Disadvantages:
- Loss of heat through thermocouple wires may lead to error in measured temperature.
- Resistance of thermocouple wire may affect emf displayed on meter.
- Response may change with time due to the diffusion of impurities.
- Limited range of linearity.
- Accuracy limited to about 1%.

How it works:

1. Consider a single length of metallic conductor where the temperature of one end is raised relative to the other. The number density of mobile electrons increases with increasing temperature and leads to a concentration gradient of electrons between the hot and cold ends of the conductor. Due to this gradient, diffusion of electrons occurs from the hot end to the cold end. The hot end becomes positively charged. This is the **Thomson effect**.

2. Now consider two lengths of dissimilar metals joined at one end. There is a difference in the density of electrons in the two materials. Thus, there is a concentration gradient of electrons at the junction which results in diffusion of electrons across the junction. This diffusion means that the material with the higher density of electrons becomes positively charged. This is the **Seebeck effect**.

These two effects lead to a **contact potential** at the junction of two dissimilar metals, the magnitude of which depends upon the temperature and the nature of the metals. The difference in contact potentials between the two junctions is a measure of the temperature difference between them.

The selection of metals which are used to make thermocouples depends upon the range of temperatures to be measured. The **thermoelectric sensitivity** ($\mu V/°C$) of a particular material is stated with respect to platinum at 0 °C.

Material	Sensitivity ($\mu V/°C$)
Constantan*	−35
Copper	+6.5
Iron	+18.5
Platinum	0

The introduction of a third metal into the thermocouple circuit does not alter the difference in **contact potentials** between its ends as long as the newly introduced junctions are both at the same temperature. This means that the ends of a thermocouple may be brazed or soldered together without affecting the operation of the device.

A commonly used thermocouple is copper/constantan. The sensitivity is thus: $+6.5 − (−35) = 41.5 \ \mu V/°C$.

Several standard pairs of materials are in common use and are conventionally given character labels, e.g. "Type K".

Temperature range °C

		−200	+300	4.24 mV
Copper/constantan*	T	−200	+300	4.24 mV
Iron/constantan	J	−200	+1100	5.268
Chromel/alumel*	K	−200	+1200	4.10

* Constantan is an alloy of 60%Cu and 40%Ni.
  Chromel is an alloy of nickel and chromium and alumel is an alloy of nickel and aluminium

The emf produced for a hot junction at 100 °C and cold junction at 0 °C.

Thermocouples are usually **non-linear**. The output may be linearised in software using data from calibration reference tables that are available which give temperature and voltage relationships referenced to 0 °C. However, the cold junction in an actual thermocouple is usually at room temperature. **Cold junction compensation** is required to correct for this. For example, an LM335 precision temperature sensor, a solid state device which acts like a **zener diode**, can be used to offset the thermocouple voltage. The reverse bias breakdown voltage of this device is linearly dependent upon the absolute temperature and is directly calibrated in K. The thermocouple voltage corresponding to the separately measured room temperature is added to the voltage from the thermocouple and then the calibration look-up table is applied to determine the temperature at the sensor end of the thermocouple. Cold junction compensation can be done electrically in **hardware**, or by a **software** correction to the data.

1.2.8 Thermistors

Thermistors are resistive temperature elements made from **semiconductor** materials. The resistance of these elements decreases with increasing temperature (negative temperature coefficient). The correspondence between resistance and temperature is highly **non-linear**.

Advantages:
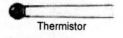
Thermistor

Some thermistors have a positive temperature coefficient.

- Inexpensive.
- Small size.
- Low mass (small time constant).
- Large output signal (high sensitivity).

Disadvantages:
- Accuracy generally not as good as Pt resistance thermometer.
- Limited temperature range: −100 to 450 °C.
- Non-linear response.
- Tolerance only about ±5%.

The relationship between resistance and temperature is exponential and has the form:

$$R_T = A e^{\frac{\beta}{T}}$$

A and β are constants which depend upon the material with which the thermistor is made. T is the temperature in K, and R_T the resistance at temperature T.

If a reference temperature T_0 is chosen, then this equation can be expressed as:

$$R_T = R_0 \exp\beta\left(\frac{1}{T} - \frac{1}{T_0}\right)$$

where R_0 is the resistance at T_0, usually taken to be 25 °C.

A typical value of resistance at room temperature is 10 kΩ falling to about 1 kΩ at 100 °C.

1.2.9 Relative humidity

The traditional method of
measuring humidity is by the use
of a wet and dry bulb
psychrometer. In this device, a
wick, soaked in distilled water, is
placed over the bulb of a mercury-
in-glass thermometer. Another
identical thermometer is placed
nearby with nothing over the bulb.
The air whose relative humidity is
to be measured is blown over the
bulbs of both thermometers.

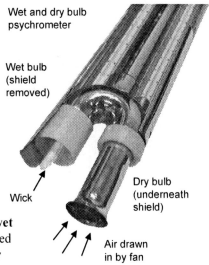

Wet and dry bulb
psychrometer

Wet bulb
(shield
removed)

Wick

Dry bulb
(underneath
shield)

Air drawn
in by fan

The evaporation of water from the **wet
bulb** causes the temperature measured
to fall compared with that of the **dry
bulb**. A psychrometric chart is used to
read off the relative humidity from the
wet and dry bulb thermometer
readings.

Relative humidity can be measured
electronically. One device uses the
change in capacitance between two
gold films separated by a **mylar sheet**.
As water is **absorbed** into the mylar,
the capacitance changes and this can be
measured electronically. In another
device, the change in **capacitance** of
two **silicon wafers** on opposite sides of
a glass slide is measured. The
capacitance depends upon the relative
humidity of the air surrounding the
device. A temperature sensor mounted
above the device is used to compensate
for differences in response at different
ambient temperatures.

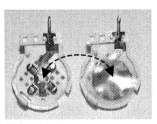

Mylar sheet type sensor

Silicon wafer type sensor

1.2.10 Review questions

1. A Pt resistance thermometer is to be used to measure
    temperature. The relationship between resistance and
    temperature is to be given by the following equation:

    $$R_{Pt} = R_o\left(1 + AT + BT^2\right)$$

    If $R_o = 100\ \Omega$, $R_{100} = 138.50\ \Omega$, and $R_{200} = 175.83\ \Omega$, determine:
    (a) the value of the constants A and B;
    (b) the fundamental interval.

2. A Pt resistance element is marked as per the following diagram. What
    is the function of each pair of terminals?

3. BS1904 specifies that the fundamental interval for a Pt resistance
    thermometer should be $38.5\ \Omega$. What does the term "fundamental
    interval" refer to?

4. A mercury-in-glass thermometer is marked as follows:

    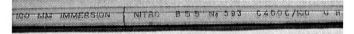

    Identify the meaning of each of these markings.

5. A mercury-in-glass thermometer made to an approved standard contains
    a widening of the capillary tube at the top of the instrument. What is the
    purpose of this widening and why must it have a spherical top?

6. An optical pyrometer uses a disappearing filament to enable an estimate
    of temperature to be made. In what way does the filament disappear
    and what is the significance of the disappearance?

7. Discuss the relative merits of the arrangement of thermocouple connections as shown:

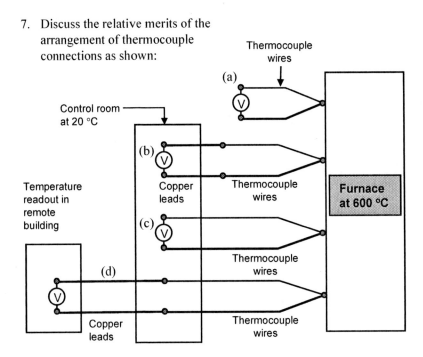

8. Refer to an iron/constantan thermocouple (Type J) table.

(a) If the cold junction is at a room temperature of 25 °C, and the reading on the millivolt meter is 24.4 mV, determine the temperature of the hot junction.

(b) The calibration table for an iron/constantan thermocouple can be approximated by the following formula:

$$Emf = 0.05038T + 0.3047 \times 10^{-4} T^2$$
$$- 0.8566 \times 10^{-7} T^3 + 0.1334 \times 10^{-9} T^4$$

Using the mV reading given in (a), determine the percentage difference between the results given by the table and the formula for the temperature in (a).

(Ans: 469°C, 8%)

1.2.11 Activities

A thermocouple consists of a length of two dissimilar metals which are joined at either end. One of the junctions is commonly held at a reference temperature, and the other junction is exposed to the temperature to be measured. The voltage measured across a break in one of the wires depends on the difference in temperature between the two junctions of the device.

In a typical application, one end of a thermocouple is usually brazed or soldered together to form the sensor and the other ends of the wires are connected directly to the voltmeter. In this case, the reference junction is at room temperature.

To overcome variations in voltage which would occur due to changes in room temperature, a third temperature measuring device may be employed to provide cold junction compensation. The third temperature sensor measures an absolute value of room temperature and provides a voltage, which when added to the thermocouple voltage, produces a total emf as if the cold junction of the thermocouple was at 0 °C.

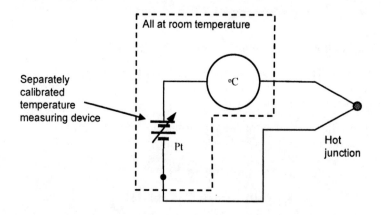

(a) Thermocouple calibration

The aim of this part of the experiment is to construct a chromel/alumel thermocouple and to calibrate it against a known standard.

1. Set up the apparatus as shown in the diagram and set the digital voltmeter to the 100 mV range.
2. Check the temperature of the ice bath at regular periods to ensure it remains at 0 °C.
3. Heat the water containing the hot junction gently and so obtain a range of temperatures between 0 °C and 100 °C with corresponding readings on the voltmeter.
4. Take sufficient readings to enable you to construct a good calibration graph of temperature (°C) vs thermocouple voltage (mV).
5. Compare your data with that given by BS4937.

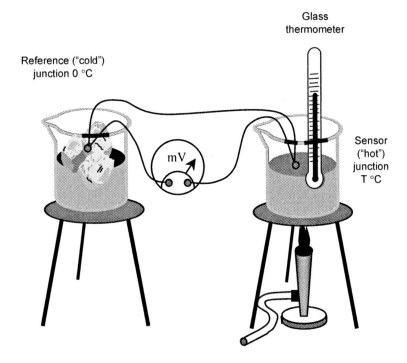

(b) Cold junction compensation

Cold junction compensation is a process whereby a voltage is added (or subtracted) from the output voltage of the thermocouple so that the reference junction appears to be at 0 °C even if it is not. This may be done electronically using an LM335 precision temperature sensor. The LM335 is a solid-state device which acts like a zener diode. The reverse bias breakdown voltage is linearly dependent upon the absolute temperature and is directly calibrated in K. For example, the output from the LM335 extrapolates to 0 V for 0 K. At 273 K, the breakdown voltage is 2.73 V.

Errors due to self-heating of the device can be minimised by selecting a current limiting resistor R_1 to reduce the operating current through the device to a minimum – enough to drive the device into breakdown at maximum temperature for the application as well as any externally applied loads. A current of about 1 mA is reasonable.

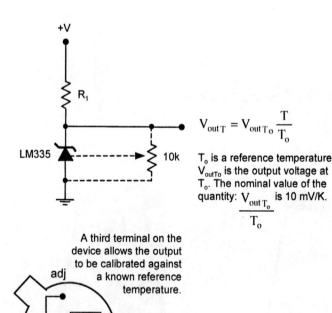

$$V_{out\,T} = V_{out\,T_0}\frac{T}{T_0}$$

T_0 is a reference temperature V_{outT_0} is the output voltage at T_0. The nominal value of the quantity: $\dfrac{V_{out\,T_0}}{T_0}$ is 10 mV/K.

A third terminal on the device allows the output to be calibrated against a known reference temperature.

adj

LM335H

Now, the thermocouple output voltage is determined by the temperature difference between the hot and cold junctions and the relative thermoelectric sensitivity of the metals. The Seebeck coefficient α for some common thermocouples is shown below:

Copper/constantan	T	$\alpha = 42.4\ \mu V/K$
Iron/constantan	J	52.8
Chromel/alumel	K	41.0

The voltage output from the LM335 has to be matched with the voltage range of the thermocouple material and this can be done using a simple voltage divider resistor network.

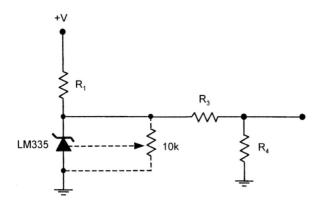

The nominal sensitivity of the LM335 is 10 mV/K and we need to factor this down to the $\mu V/K$ range. We choose R_3 and R_4 so that:

$$R_4 = \frac{\alpha}{10mV} R_3$$

For example, if $R_3 = 220\ k\Omega$, and we are using a Type K thermocouple, then R_4 is:

$$R_4 = \frac{40.8 \times 10^{-6}}{10 \times 10^{-3}} 220,000$$

$$= 898\ \Omega$$

The LM335 is calibrated in K. However, the voltage output from the thermocouple must be referenced to 0 °C = 273 K.

At 273 K, the output or reverse bias breakdown voltage of the LM335 will be about +2.73 V. This voltage, factored down to match the thermocouple, must then be applied to the negative side of the thermocouple to raise its potential. Thus, when the hot junction is at 0 °C, there will be no voltage difference between the output terminals.

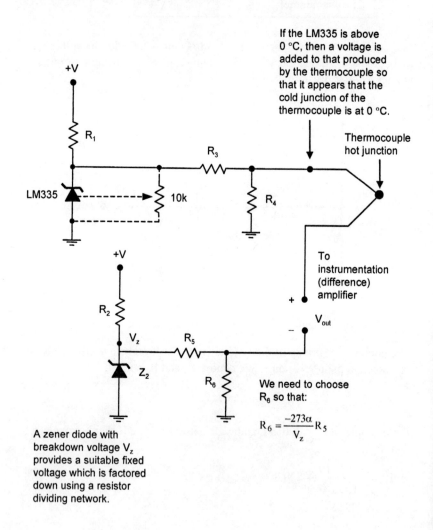

If the LM335 is above 0 °C, then a voltage is added to that produced by the thermocouple so that it appears that the cold junction of the thermocouple is at 0 °C.

Thermocouple hot junction

To instrumentation (difference) amplifier

V_{out}

We need to choose R_6 so that:

$$R_6 = \frac{-273\alpha}{V_z} R_5$$

A zener diode with breakdown voltage V_z provides a suitable fixed voltage which is factored down using a resistor dividing network.

1. With $+V = 5$ V, calculate a suitable value of R_1 to limit the current through the LM335 to 1 mA.
2. Calculate a suitable value of R_4 with $R_3 = 220$ K to match the characteristics of the supplied thermocouple material.
3. Calculate a suitable value of R_2 so that the maximum current through the zener diode Z_2 is no more than 1 mA with a $+5$ V supply.
4. Calculate suitable values of R_5 and R_6 so that the appropriate "zero" offset is applied to the negative end of the thermocouple (select $R_5 = 220$ k).

Parts list:
1x LM335H precision temperature reference
1x IN4732 4.8 V zener diode
2x 220 k; 1x 2.2 k; 1x 100 Ω; 1x 4.7 k; 1x 680 Ω; 1x 1 k

Data acquisition system

In order to interface our thermocouple to the serial data acquisition system presented in Part 2 of this book, we need to amplify the signal to obtain an appreciable voltage for subsequent conversion to digital format. The full-scale output from the thermocouple circuit depends of course on the maximum temperature being measured. Let us work with the range 0 to 100 ºC. At 100 ºC, the output from a K type thermocouple (including any cold junction compensation) will be 4.1 mV. A convenient voltage gain required would thus be about 1000 to give an analog input of around 4 V to the ADC. The overall aims of the project activity in this book are to:

1. Design, build and test a thermocouple circuit which employs cold junction compensation. The output of the circuit is to be a differential voltage which is proportional to the temperature of the hot junction of the thermocouple over the range 0–100 ºC.

2. Construct an instrumentation amplifier which will take in the voltage levels from a cold-compensated thermocouple circuit and provide a 0–5 V output voltage which is proportional to the temperature of the hot junction of the thermocouple (ºC).

3. Using the serial data acquisition system, construct a computerised temperature recording and control system using the software of your choice.

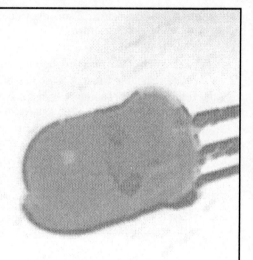

1.3 Light

1.3.1 Light

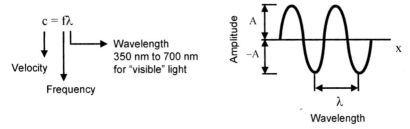

Light is an electromagnetic wave.

The velocity of light waves in a vacuum is $c = 2.99792458 \times 10^8$ ms^{-1}

For any type of wave, including light waves:

For visible light, the frequency is of the order of 10^{15} Hz. Photodetectors cannot respond to such rapid changes and thus generally indicate rms or mean values of power of the radiation.

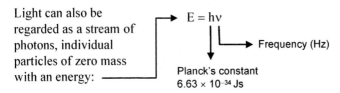

The energy of a photon depends on the frequency but is in the order of 10^{-20} J. Most detectors respond to large numbers of photons but there are a few that can provide a useful output for just one photon.

There is a difference between the measurement of radiant energy and that of light. The term light implies a human connection, and the measurement of *visible* light is called **photometry**. The human eye is a very common photometric detector.

Radiometry is concerned with the measurement of radiant energy independent of the type of detector used. A third field of radiation measurement is concerned with the quantum nature of light and is called **actinometry** – which arose from a study of the photochemical effects of light.

1.3.2 Measuring light

Many units concerning light are in daily use but before we introduce them, we need to consider the definition of a solid angle, the **steradian** (sr). ➤

The steradian Ω is the solid angle which, having its vertex in the centre of a sphere, cuts off an area of the surface of the sphere equal to that of a square with sides of length equal to the radius of the sphere.

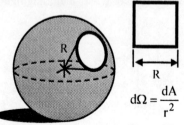

$$d\Omega = \frac{dA}{r^2}$$

The solid angle is the surface area of the cone divided by the square of the radius.

Radiometric definitions

Radiant energy is energy received or transmitted in the form of electromagnetic waves and has the units of joules (J).

Radiant power or flux is radiant energy received or transmitted per unit time in watts (W).

Radiant flux density (*irradiance*) is the radiant power incident on a perpendicular unit area and has the units (Wm^{-2}).

Radiant intensity is the radiant power per unit of solid angle (W/sr).

Actinometric definition

Photon flux is the actinometric equivalent of radiant flux and is the number of photons impinging on a surface per second. Each photon has an energy = hv.

Photometric definitions

The unit of **luminous flux** is the **lumen** (lm = cd.sr). Luminous flux is the radiant flux weighted by a spectral efficiency factor which characterises the response of the human eye. The lumen is the human equivalent of radiometric power (W).

The brightness of a surface is called the **illuminance** and has the unit **lux** and is the luminous flux per unit area (lx = 1 lm/m²). It is the human or photometric equivalent of radiant flux density.

The luminous flux per solid angle is called the **luminous intensity** and is given the unit **candela** (cd). It is the photometric equivalent to radiant intensity.

1.3.3 Standards of measurement

The official SI base unit for measuring the luminous intensity of light is the **candela**. The candela is the only SI base unit which has its origins in the response of a human organ (i.e. the eye) – it is a photometric quantity. The candela is a base SI unit upon which lumens and lux are derived.

The candela has become an important base unit due to the historical nature of measurements of light which involved the human eye as the detector. Early standards by which the response of a human eye were quantified involved candles, flames, and incandescent lamps. Human observers compared an unknown light source to a standard.

Modern methods utilise the response of a device (e.g. a photocell) which has spectral characteristics which are very close to that of a **standard observer**. Standard sources provide a way of calibrating photocells to be used in industry.

> The **candela** is defined as the luminous intensity in a given direction of a source that emits monochromatic radiation of frequency 540×10^{12} Hz and has a radiant intensity of $1/683$ W/sr in that direction.

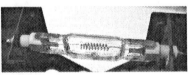

Standard light source

For precise photometric work, it is usually preferable to operate lamps on DC. It is preferable to set the operating current. Measurement of luminous flux may be made by comparison with luminous flux standards using an integrating sphere, or a goniophotometer.

Standard photometer

The most commonly used measurement of light intensity is not actually the candela, but the **illuminance** or **brightness** and is typically given in **lux**. In a normal lecture room, the illuminance is about 300 lux. A bright summer's day: 20000 lux. In daylight, 680 lux corresponds to a radiant flux of about 1 Wm^{-2}

1.3.4 Thermal detectors

In thermal detectors, incoming radiation results in a change of temperature of the sensor. The temperature of sensor is an indication of the magnitude of incident radiation.

Temperature is usually measured with a **thermopile**, which consists of a large number of thermocouples in series. The sensitivity of a thermal detector using a thermopile with a surface area of 1–10 mm^2 is typically about 10–100 V/W, with a time constant of about 10 ms.

The sensitive region of the detector is usually blackened so as to absorb the maximum amount of incoming radiation.

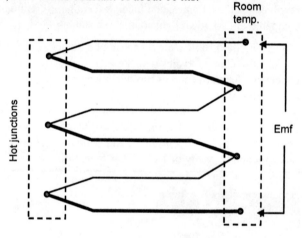

Another type of thermal detector employs thermistors instead of a thermopile to measure temperature. Such a device is called a **bolometer**.

Still yet another type of thermal detector utilises the **pyroelectric** property of certain ferroelectric materials. Incident radiation causes a change in the surface charge of a residually polarised ceramic. The effect can only be measured in a pulsed mode of operation and hence an AC amplifier is used to produce a reasonable output.

1.3.5 Light dependent resistor (LDR)

In a semiconductor, **photoconductivity** is a result of an increase in
electrical conductivity due to impingement of photons on the
semiconductor material. This increase can only occur if the incident
photons have an energy $hv > E_g$ where E_g is the energy gap between the
valence and conduction bands.

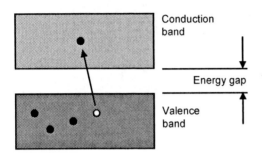

Conduction
band

Energy gap

Valence
band

Incident photons cause
electrons in the valence
band to be given energy
hv and, if $hv > E_g$,
valence electrons enter
the conduction band,
leading to an increase in
the number of mobile
electrons.

The increase in conductivity manifests itself as an increase in the current
through the device for a given applied voltage and as such may be called a
light dependent resistor (LDR).

When an LDR is illuminated with a steady beam, an equilibrium is
reached where the decay of electrons is matched by the excitation.

The ratio of the number of excited electrons to the number of incoming
photons is called the **quantum efficiency** and is dependent on the
probability of the number of elastic collisions between photons and
electrons.

For a given frequency of incident beam, the number of mobile electrons
created is a function of intensity of the beam. However, the conductivity of
the material depends not only on the intensity of the incident radiation, but
also upon its **frequency**. This is due to the filling of available quantum
states.

LDR

e.g. A popular material for LDRs
is cadmium sulphide (CdS). CdS
has a peak response at 600 nm,
$E_g = 2$ eV and matches the
frequency response of the human
eye quite closely. In contrast, lead
sulphide (PbS) ($E_g = 0.4$ eV) has a
peak response at 300 nm.

1.3.6 Photodiode

A photodiode employs the **photovoltaic** effect to produce an electric current which is a measure of the intensity of incident radiation.

1. Near the junction, concentration gradient causes free electrons from n side to diffuse across junction to p side and holes from p side to diffuse across to n side.

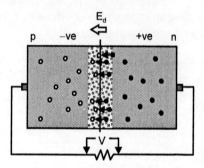

2. Resulting build-up of negative charge on p side and positive charge on n side establishes an increasing electric field E_d across the junction.

Current will flow in external circuit as long as photons of sufficient energy strike the material in the depletion region.

The area near the junction becomes free of majority carriers and is called the **depletion region**. When a photon creates an electron-hole pair in the depletion region, the resulting free electron is swept across the junction towards the n side (opposite direction of E_d).

Even though the photodiode generates a signal in the absence of any external power supply, it is usually operated with a small reverse bias voltage. The incident photons thus cause an increase in the reverse bias leakage current I_o.

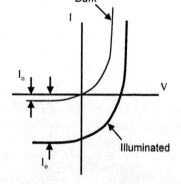

Photodiode

The reverse bias current is directly proportional to the luminous intensity. Sensitivity is in the order of 0.5 A/W.

1.3.7 Other semiconductor photodetectors

Avalanche photodiodes operate in reverse bias at a voltage near to the break-down voltage. Thus, a large number of electron-hole pairs are produced for one incident photon in the depletion region (internal ionisation).

Phototransistors provide current amplification within the structure of the device. Incident light is caused to fall upon the reverse-biased collector-base junction. The base is usually not connected externally and thus the devices usually only have two pins. Increasing the light level is the same as increasing the base current in a normal transistor.

Schottky photodiodes use electrons freed by incident light at a metal–semiconductor junction. A thin film is evaporated onto a semiconductor substrate. The action is similar to a normal photodiode but the metal film used may be constructed so as to respond to short wavelength blue or ultraviolet light only since only relatively high energy photons can penetrate the metal film and affect the junction.

PIN photodiode is a pn junction with a narrow region of intrinsic semiconductor sandwiched between the p and n type material. This insertion widens the depletion layer thus reducing the junction capacitance and the time constant of the device – important for digital signal transmission via optical cable.

A **charge coupled device** CCD is an array of closely spaced photodiodes. Incident light is converted to an electric charge in each diode. A sequence of clock pulses transfers the accumulated charge to a digital output stream. For video applications, an image must be focussed on the device using a lens.

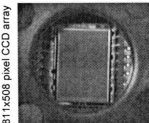

811x508 pixel CCD array

1.3.8 Optical detectors

Optical detectors are characterised by:

Responsivity/Sensitivity $\quad \dfrac{P_{out}}{P_s}$

Output power $\nearrow$ P_{out}

Input (signal) power

Spectral response $\qquad$ Sensitivity as a function of incident wavelength

Detectivity D $\qquad D = \dfrac{SNR}{P_s} \quad W^{-1}$

Signal to noise ratio

Input power

Detectivity D*

Independent of area and bandwidth.

$$D* = D\sqrt{A\Delta f} \quad cm^{1/2}Hz^{1/2}W^{-1}$$

Bandwidth

Detector area

Detectivity D**

Independent of field of view.

$$D*(\theta) = D* \dfrac{2\pi}{\sin\theta}$$

Half angle

Quantum efficiency η

$$i_s = \left(\eta\dfrac{P_s}{h\nu}\right)e$$

Charge on electron

Signal current

Photons/sec

1.3.9 Photomultiplier

One of the most common applications of photomultipliers is for the detection of nuclear radiation. But, the device may be also used as the basis for detection of a wide range of phenomena which involve very low light output levels (e.g. chemi-luminescent gas detector).

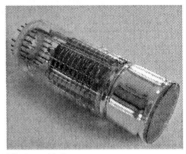

The light sensitive surface of a photomultiplier consists of a thin film of an alkali metal which has a low **work function** W. When a photon with energy E impinges on the metal, if E > W, then electrons are emitted from the metal. These electrons are accelerated by an applied potential (of about 200 V) towards a **dynode**. An accelerating electron, when it strikes the dynode, has sufficient kinetic energy to eject two or more electrons from the dynode material.

These two electrons are then accelerated through another 200 V potential to another dynode and thus cause four electrons to be ejected. This **amplification** may involve several stages of dynodes, each at a potential of 100–200 V above the previous stage. Thus, the final electron current is sufficiently high to measure with conventional electronic equipment.

Amplification is thus done within the evacuated structure and may be as high as 10^6. Further, this amplification is done prior to the intrumentation amplifier input resistance and noise in the signal is thus reduced considerably. Dark current (due to thermionic emission at cathode) limits detectivity.

Note: A high voltage power supply is needed to produce the required accelerating potentials at each dynode.

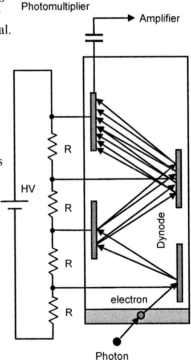

Photomultiplier

Amplifier

HV

R

R

R

R

Dynode

electron

Photon

1.3.10 Review questions

1. What is the photon flux incident on a 1 m^2 surface being illuminated by 60 W of light of wavelength 620 nm.   (Ans: 1.87 × 10^{20}/sec)

2. A 100 W motor cycle headlamp can just be seen by a pedestrian two kilometres away. The size of the pupil in the pedestrian's eye is 1 mm^2. Calculate the minimum incident power detectable by the retina of the eye. Assume that the headlamp is 25% efficient in converting electrical energy into visible light.   (Ans: 0.5 × 10^{-12} W)

3. Discuss the differences between the radiometric and photometric definitions of light. That is, why are they different?

4. A photodiode has a sensitivity of 9 nA/lux at 560 nm and an area of 40 mm^2. Express the detectivity in A/W. Note: 1 lux = 1 lumen/m^2. A radiant flux of 1 W at 560 nm produces 685 lumens.

   (Ans: 0.154 A/W)

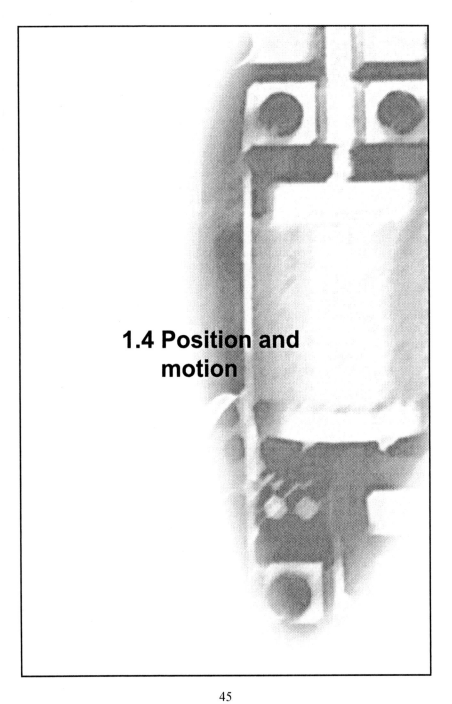

1.4 Position and motion

1.4.1 Mechanical switch

A simple contact type transducer
converts displacement into an electrical
signal and may take the form of a
mechanically operated **switch**.

Switches are specified as single or
double pole (one or two rows of parallel
contacts), single or double throw (centre
contact switches from one contact to
another). The size of the contact pads
depends upon the current and the type of
load in the circuit. For **high voltage**
switching, the contacts are immersed in
oil to reduce the occurrence of **arcing**.

Microswitch

Contact points

The main problem with switches in
interfacing applications is that of **bounce**.
Most switches contain a spring to keep the
contacts either together or apart. When the
switch is closed, the spring often causes
the contacts to bounce, thus creating a
series of make and break contacts over a
period of a few milliseconds. If any
interfacing circuit should be monitoring the
switch, then it might register the opening
and closing of the switch during bounce
contact. To avoid this, the interfacing
system needs to incorporate switch
debounce circuitry or software logic.

In software debouncing, the
program may wait for 10 or 20
msec after first registering an event
and test the switch again before
proceeding.

In hardware debouncing, the
output of the switch can be
processed by a latch circuit. The
output of the latch will only change
state if the inputs change by TTL
level signals.

Wear of the contact points in a switch
occurs mainly due to **arcing**. This is
especially important when a switch is
used across an inductive load. Opening
such a switch causes a very high voltage
to be induced across it (due to Lenz's law)
which leads to arcing across the gap. For
this reason, some switches have a gap and
opening rate specifically designed for DC
and AC applications.

Electrons are ejected from the
negative (or cathode) side of the
switch and accelerated towards the
positive side (anode). This causes
ions to be dislodged from the
anode and be accelerated towards
the cathode. Material accumulates
on the cathode and cavities appear
on the anode.

1.4.2 Potentiometric sensor

A **potentiometric sensor** converts a linear or angular displacement into a change in resistance. The sensor itself may be made from a coil of wire over which a moving contact or wiper causes a change in resistance between the terminals of the device.

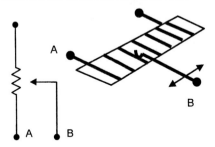

A coil of wire is wound on a mandrel. If the winding is uniform and the wire is of a constant cross-section A and resistivity ρ, then the resistance R is:

$$R = \rho \frac{l}{A}$$

l is the total length of wire between A and B.

A very common application is the **fuel level sensor** in a motor vehicle. The sensor adjusts the resistance between its terminals according to the level of fuel in the fuel tank and thus indicates the displacement of the surface of the liquid fuel as fuel is consumed by the engine.

A non-linear response can easily be obtained by altering the dimensions of the **mandrel**. For example, to show a larger deviation in R at low fuel level (increased sensitivity), then the mandrel can be shaped or the spacing of the windings altered according to position.

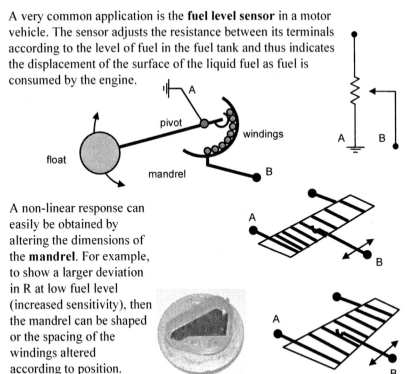

Resistance mandrel

1.4.3 Capacitive transducer

A capacitive sensor converts a change in position or change in properties of the **dielectric** material into an electrical signal.

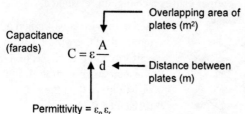

Capacitance (farads)

$$C = \varepsilon \frac{A}{d}$$

Overlapping area of plates (m²)

Distance between plates (m)

Permittivity = $\varepsilon_0 \varepsilon_r$

Alteration of any of these three parameters leads to a change in capacitance which may be measured electrically.

ε_r is the relative permittivity of the dielectric

Permittivity of free space $\varepsilon_0 = 8.85 \times 10^{-12}$ Fm⁻¹

Examples:

1. Overlapping area of semicircular plates alters with **angular displacement** of shaft.

   The capacitance is proportional to the angular displacement. Let A_0 = area of plate at $\theta = 0$. The overlapping area A is computed from:

   $$A = A_0 \frac{(180 - \theta)}{180}$$

   The capacitance is thus:

   $$C = \frac{\varepsilon A_0}{d} \frac{(180 - \theta)}{180}$$

   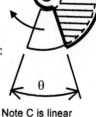

   The sensitivity is dC/dθ:

   $$\frac{dC}{d\theta} = \frac{-\varepsilon A_0 (N-1)}{180d}$$

   N is total number of moving and stationary plates

   Note C is linear w.r.t. θ

   Farads/degree

2. Change in **dielectric property** of material between plates can also be used.

   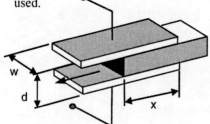

   The sensitivity is dC/dx:

   $$C = \varepsilon_0 \frac{A}{d} + (\varepsilon - \varepsilon_0) \frac{w}{d} x$$

   $$\frac{dC}{dx} = (\varepsilon - \varepsilon_0) \frac{w}{d} \qquad \text{farads/m}$$

1.4.4 LVDT

The most commonly used inductive transducer is the linear variable
differential transformer (**LVDT**). In this device, a **core** is mounted on a rod
which passes through the centre of a coil and which is connected to the part
to be moved. Changes in the **magnetic coupling** between the coils convert
a mechanical movement into an electrical output.

The diagram shows the primary
coil being driven by an
oscillator. If the core moves
upwards the flux linkage to the
upper coil is increased and V_1
increases. The magnetic flux
linkage to the bottom coil
decreases and V_2 thus decreases.
The displacement of the core is
thus registered as $(V_2 - V_1)$.

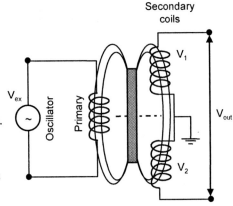

The output voltage also depends
on the **driving frequency** and
voltage amplitude V_{ex}.

When the core is at the central or
null position, the output voltage is
zero. As the core is moved above
and below the null position, the
output signal rises and falls the
same amount but undergoes a
change of **phase** by 180°.

Note: Arrangement of secondary coils
means that the voltage induced in each of
them is opposite in polarity:

$$\Delta V_{out} = \Delta V_1 - \Delta V_2$$

In order to extract the sign (and
therefore the direction of motion),
it is necessary to use a synchronous
demodulation technique. Dedicated
ICs such as the AD698 or NE5521
can be used for LVDT drive and
signal processing functions.

LVDT core and coil

The sensitivity of an LVDT is specified in mV/mm/V_{ex}. Typical range of
displacements is ±0.25 mm to ±250 mm. Typical drive frequency of V_{ex} is
about 1–10 kHz. With proper instrumentation, an LVDT can resolve less
than a **nanometre** of movement.

1.4.5 Angular velocity transducer

Electromagnetic induction used to produce a voltage which depends on
the velocity of a coil which moves relative to a fixed magnet (or vice
versa). Some examples are:

Toothed-rotor magnetic tachometer

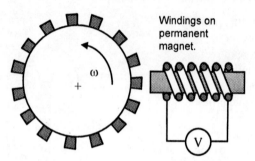

Windings on
permanent
magnet.

Magnetic teeth on rotor modifies the **magnetic circuit** when the rotor is rotating. This induces a voltage in the windings which surround the magnet.	Amplitude and frequency of the output voltage are directly proportional to the rotational speed ω.

Drag-torque tachometer (motor vehicle speedometer)

A permanent magnet revolves on a shaft and induces **eddy currents** in the
disk. These eddy currents themselves produce a magnetic field which
interacts with the rotating magnetic field on the rotor. The net result is a
drag force on the disk which is proportional to the speed of rotation of the
rotor.

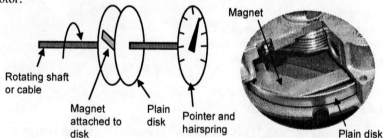

Rotating shaft
or cable

Magnet
attached to
disk

Plain
disk

Pointer and
hairspring

Magnet

Plain disk

The disk is connected to a pointer and a hairspring. The scale is calibrated
to indicate velocity in the desired units (e.g. mph or km/h).

1.4.6 Position sensitive diode array

A **diode array** is an assembly of 1024 individual photodiodes in a linear array. The device is particularly useful for **spectrophotometer** applications where light, spread by a prism, is shone onto the array and the intensity of the wavelength spectrum measured simultaneously. In X-ray absorption spectroscopy and **xray diffraction** a diode array is used as a **position sensitive detector (psd)** to determine the angle of diffraction of an xray beam.

The output from a diode array is an analog signal that gives the distance from the edge of the array to the **centroid** of the incident light spot. The response of a diode array is very fast (**rise time** $\approx$ 5 μsec) and the device has very high **positional resolution** ($\approx$250 nm) and a linearity of less than 1% of full scale.

25 mm

In the array assembly, a resistance material is placed on one side of a pn junction. Light impinging on the junction (held in reverse bias with about 12 V) generates a current whose maximum value is at the centroid of the greatest power density of the incident light. This current flows along the resistance material to the connecting electrode. The output current signal thus depends on the **total resistance** from the electrode to the spot at which the current is generated.

Ambient light will cause a signal to be generated in the device corresponding to the centre of the array. The spot size for the position being measured should be made as bright as possible without damaging the photodiodes by heating them excessively. It should be noted that the device will only respond to the distribution of light that actually falls on the sensitive elements of the array and so the output reflects the position of the centroid of the spot received by the array – which might not be the same as that incident on the device as a whole for a large incident spot.

1.4.7 Motion control

Positional encoders are usually fitted to motion control systems to provide position and velocity feedback to a PID controller to control motion. The PID controller in turn generates a voltage signal that produces a velocity profile that will ensure that the motion is accomplished as desired.

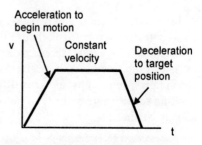

Rotary encoder

A rotary encoder in its most simple form comprises a disk in which there are slots at precise regular intervals. The disk is typically mounted on a shaft whose rotation is to be measured. The shaft can in turn be part of a thread with a zero backlash ball nut that transfers rotary motion into linear motion. The movement of the disk is measured by a photocell that detects light from an emitter on the other side of the disk. The resolution or **step size** is the angle between the slots.

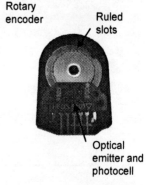

Linear encoder

A linear track encoder counts the number of lines over which is moved an optical emitter and photocell. This has the advantage of a linear movement being a measure of the actual distance moved rather than the rotation of a geared screw thread. The accuracy depends upon that of the ruled lines on the track which are usually in the order of 20 μm spacing.

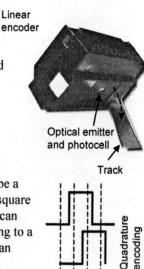

The output signal from an optical encoder may be a **quadrature signal**. The encoder produces two square waves out of phase by 90°. A motion controller can then extract four **phase changes** per cycle leading to a four-fold increase in step size. Some encoders can further **interpolate** the signals from the slots or gratings by up to a factor of 50. Step sizes of 0.1 μm are routinely available with these types of encoders.

1.4.9 Review questions

1. In a capacitance transducer, the capacitance is usually connected to the input of a charge amplifier. This type of amplifier has a gain which is not dependent on frequency. Why would this be an advantage?

2. A variable dielectric capacitive displacement transducer sensor consists of two square metal plates of sides $w = L = 5$ cm separated by a gap of 1 mm. A sheet of dielectric material 1 mm thick and the same area as the plates can be slid between them.

   (a) If the dielectric constant of air is $\varepsilon_r = 1$ and that of the dielectric material is $\varepsilon_r = 4$, calculate the capacitance of the sensor when the input displacement $L - x = 2.0$ cm.
   (b) Determine the sensitivity of the device.

   Hint: $C = \varepsilon_r \varepsilon_0 \dfrac{A}{d}$

   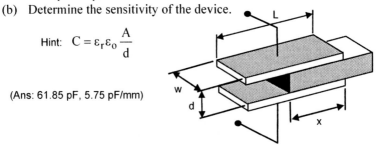

   (Ans: 61.85 pF, 5.75 pF/mm)

3. A motorised specimen positioning table has a rotary encoder with a line count of 2000. It is attached to a lead screw of pitch 2 mm which translates rotary motion into a linear motion. Calculate the linear step size (in μm) for the device.   (Ans: 1 μm)

4. A switch is in the process of opening and the open circuit voltage is 12 V. Compute the electric field strength when the gap between the contacts is 1 μm.   (Ans: 1.2×10^7 V/m)

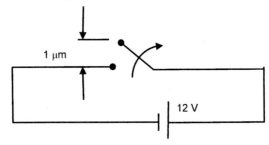

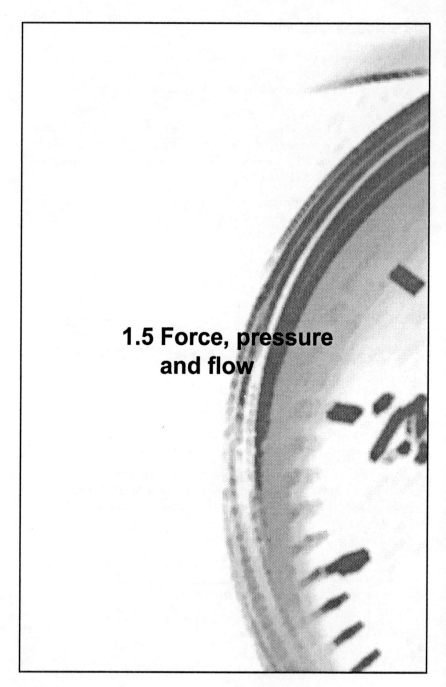

1.5 Force, pressure and flow

1.5.1 Strain gauge

A **strain gauge** is a metal or semiconductor whose resistance changes markedly when it is deformed. The deformation is usually taken to be a measurement of strain, and hence force, applied to a structure.

The resistance of a specimen of material of length l and cross-sectional area A is given by:

$$R = \rho\frac{l}{A}$$ ρ is the resistivity

A change in length or area with strain produces a change in resistance of a particular element.

If the resistance of a particular element is R_0 with no strain, then the **strain gauge factor** G is given by:

$$G = \frac{\Delta R}{R_0}\frac{L}{\Delta L}$$

Relative change in resistance ⟶
← Relative change in length (strain)

Strain gauges typically carry only a small current (15 – 100 mA) to avoid self-heating changes in resistance and thermal expansion errors.

The strain gauge element typically forms one leg of a bridge circuit which is used to measure changes in resistance of the device.

The sensitivity of the strain gauge measurement is dependent on the number of active arms in the bridge. Strain gauges are available with nominal **resistances** from 30 to 3000 Ω. The most common values are 120, 350 and 1000 Ω.

Strain gauge materials are selected so that changes in resistivity with strain (**piezoresistive** effect) are small, and the geometry is such that the application of strain results in a large change in length of the element. Gauge factors of about 2 are common.

Strain gauges are commonly purchased as a **metal foil** bonded onto a plastic adhesive film. The film is attached to the structure whose strain is to be measured with the "active" axis of the gauge aligned with the direction of the expected strain. Several gauges can be accommodated in the one film, each oriented in a different direction, to form a strain gauge **rosette**.

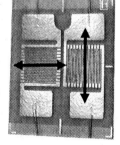

Active axes

The output from a strain gauge element will typically respond to changes in dimension arising from changes in temperature. The **thermal expansion** properties of strain gauge material are usually matched to suit the specimen material.

In one configuration, a strain gauge element is put into one arm of a bridge circuit. The **output voltage** is given by:

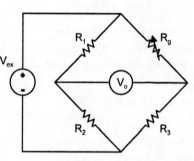

$$V_o = \left[\frac{R_3}{R_3 + R_g} - \frac{R_2}{R_1 + R_2} \right] V_{ex}$$

At balance, $R_1 R_3 = R_2 R_g$ and $V_o = 0$. When the resistance R_g changes, there is a change in V_o. The change in R_g depends upon the gauge factor G and the strain ε such that:

$$\Delta R = \varepsilon R_g G$$

Letting $R_1 = R_2$ and $R_3 = R_g$, the output voltage for a change ΔR in R_g is given by:

$$V_o = \frac{1}{2} \left[\frac{1}{1 + \varepsilon G/2} - 1 \right] V_{ex}$$

$$= -\frac{\varepsilon G}{4} \frac{1}{1 + \varepsilon G/2}$$

Note that this expression gives a **non-linear output** with changing strain. Before measuring strain, the bridge must be nulled (or **zeroed**) in the absence of any strain. Strain is then applied, and the output voltage measured.

Typical strain gauge circuits use an excitation voltage of 5 – 10V. The output signal is in the mV range. Since very small output voltages are involved, the resistors that make up the bridge circuit have to be precision matched, and the excitation voltage has to be extremely stable.

Quarter bridge strain gauge circuit

Sense leads

Voltage drops caused by resistance in the wires connecting the excitation voltage to the bridge can be a source of error when the strain gauge transducer is located some distance from the signal processing circuit. A technique called remote sensing can be used to compensate for this. With feedback remote sensing, extra sense wires are connected to the connection of V_{ex} to the bridge circuit. These sense wires are used to regulate or control the voltage supply so that the required V_{ex} is obtained at the bridge. Another method uses a direct measurement of V_{ex} applied to the bridge. This measured voltage is then used as the value for V_{ex} in the calculations.

Strain gauges can be **calibrated** in the field by using a **shunt** resistor of known value which temporarily replaces a gauge in the bridge and thus simulates a known strain for the measurement electronics.

1.5.2 Force

The most common type of force transducer is the **piezoelectric** type. In this device, force is applied to a piezoelectric crystal such as **quartz** or **lead zirconate titanate (PZT)**. The force acting on the crystal displaces the atoms within it. This displacement results in a net **charge** on the opposite faces on the crystal which can then be measured electrically. The charge is directly proportional to the force.

$$q = (d)(F)$$

d is the sensitivity of the device (i.e. the amount of charge per unit of applied force). For quartz, $d = 2 \times 10^{-12}$ C/N.

In a **piezoelectric force transducer**, metal plates are bonded onto the surface of the crystal. The crystal is pre-stressed to provide the capability of both tension and compression measurements.

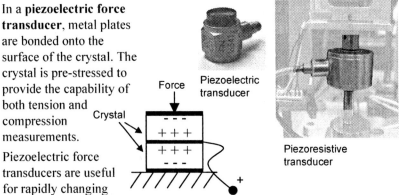

Piezoelectric transducer

Piezoresistive force transducers are useful for rapidly changing forces but their response falls off at low frequencies and are thus often termed **AC** devices. Less than a few hertz, a **DC** force type transducer is required based upon either a semiconductor or **piezoresistive strain gauge**. Deflection of the diaphragm to which the strain gauge is mounted is registered as an out of balance condition for a bridge circuit.

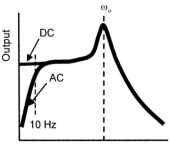

Piezoresistive transducer

At low frequencies, the **leakage** of charge on piezoelectric devices causes a reduction in signal thus limiting the frequency range to greater than a few Hertz. At higher frequencies, the charge is continually refreshed by the change in dimensions of the device and there is a linear region of operation. At frequencies larger than the **resonant frequency**, for both types of transducers, the mechanical response of the system cannot keep up with the rate of change of the applied force and again the response falls off.

1.5.3 Piezoelectric sensor instrumentation

Piezoelectric force transducers provide an output based upon the change of charge across the faces of the crystal. The capacitance of the crystal provides a voltage signal according to: $\Delta V = \Delta q/C$.

For **quartz crystals**, the capacitance is relatively low leading to a large change in voltage with a change in charge. The voltage can be measured with a **voltage amplifier** with a high input resistance (MOSFET circuitry). The **sensitivity** of the transducer is thus determined by the voltage gain of the amplifier. Such systems have a very good frequency response (1 MHz) but have a relatively high noise floor.

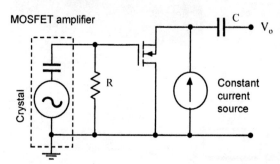

MOSFET amplifier

When the voltage across R varies, the bias voltage across the MOSFET changes and the signal is passed through the coupling capacitor C providing a measure of force.

For ceramic crystals (e.g. **PZT**) the internal capacitance is fairly high and the output can be fed into a high impedance input of a voltmeter or oscilloscope. However, the use of a **charge amplifier** allows the transducer to have a low impedance output thus making it far easier to route the signal over relatively long distances to a high impedance measurement device. By mounting the charge amplifier very close to the crystal, a very low noise output can be obtained.

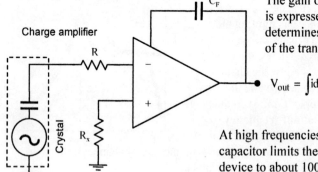

Charge amplifier

The gain of this amplifier is expressed as mV/pC and determines the sensitivity of the transducer.

$$V_{out} = \int i dt = -\frac{q}{C_F}$$

At high frequencies, the feedback capacitor limits the response of the device to about 100 kHz.

1.5.4 Acceleration and vibration

Measurement of **acceleration** is most commonly done mechanically. A **seismic mass** is supported by a spring and a **dashpot**. The mass is connected to an arm which in turn operates a piezoresistive, capacitive or inductive displacement transducer element or a **piezoelectric crystal**. The resulting output signal gives the acceleration of the reference frame of the device.

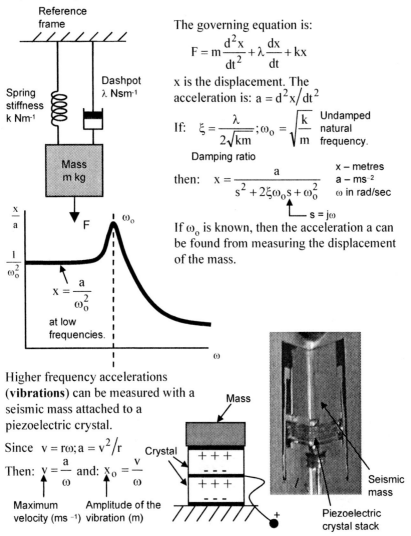

The governing equation is:

$$F = m\frac{d^2x}{dt^2} + \lambda\frac{dx}{dt} + kx$$

x is the displacement. The acceleration is: $a = d^2x/dt^2$

If: $\xi = \dfrac{\lambda}{2\sqrt{km}}$; $\omega_0 = \sqrt{\dfrac{k}{m}}$ Undamped natural frequency.

Damping ratio

then: $x = \dfrac{a}{s^2 + 2\xi\omega_0 s + \omega_0^2}$

x – metres
a – ms^{-2}
ω in rad/sec

$s = j\omega$

If ω_0 is known, then the acceleration a can be found from measuring the displacement of the mass.

Reference frame

Spring stiffness k Nm^{-1}

Dashpot λ Nsm^{-1}

Mass m kg

F

$\dfrac{x}{a}$

$\dfrac{1}{\omega_0^2}$

ω_0

$x = \dfrac{a}{\omega_0^2}$

at low frequencies.

ω

Higher frequency accelerations (**vibrations**) can be measured with a seismic mass attached to a piezoelectric crystal.

Since $v = r\omega$; $a = v^2/r$

Then: $v = \dfrac{a}{\omega}$ and: $x_0 = \dfrac{v}{\omega}$

Maximum velocity (ms^{-1})

Amplitude of the vibration (m)

Mass

Crystal

$+ + +$
$- - -$
$+ + +$
$- - -$

Seismic mass

Piezoelectric crystal stack

1.5.5 Mass

Mass may be measured by balancing the mass with a force. The force can arise from the deflection of a spring or some other linearly elastic element. In high quality balance instruments, the displacement is registered electronically and used in a feedback circuit to control a force actuator which brings the displacement back to zero. The signal to the force actuator is taken to be a measure of the mass of the specimen.

Mass balances are usually self-contained units fitted with a digital readout and a computer interface. The **computer interface** allows the balance to be configured by sending commands to it in ASCII text, often by **serial communications**. Mass values can be obtained by reading from the serial connection by sending a read command.

The format of a weighing result is usually presented in a particular format. A typical example is: | |I| | | |-|3|2|.|4|5|6| |g|CR|LF:

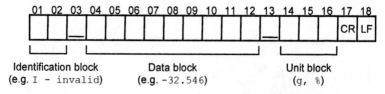

01	02	03	04	05	06	07	08	09	10	11	12	13	14	15	16	17	18
																CR	LF

Identification block   Data block   Unit block
(e.g. I - invalid)   (e.g. -32.546)   (g, %)

High quality mass balance with displacement feedback. Digital output can resolve mass increments to 0.00001 g.

1.5.6 Atmospheric pressure

Units of atmospheric pressure:

N/m^2

millibar ($=100\ N/m^2$)

mm Hg (mercury at 0 °C and at

$g = 9.80665\ ms^{-2}$)

> Note: The "torr" as used in vacuum work is 1/760 standard atmospheric pressure but is not recognised for measuring barometric pressure.
> 1 mmHg = 1 torr (named after Torricelli).

Standard atmospheric pressure is:

101.325 kPa = 760 mmHg = 29.921 inHg

Atmospheric pressure is conventionally measured with a mercury barometer since it provides a direct reading of pressure and uses a dense liquid which provides for a convenient height of the instrument.

Mercury barometers are of two general categories: Fortin and Kew. In a Fortin barometer, the mercury surface in a cistern adjusted to a fixed point prior to taking a reading – this category also includes U tube manometers.

Mercury levels are measured from top of meniscus.

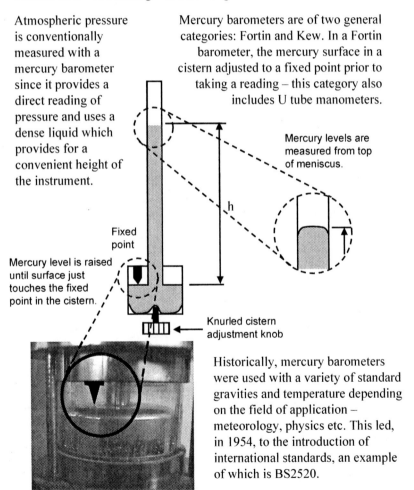

Fixed point

Mercury level is raised until surface just touches the fixed point in the cistern.

Knurled cistern adjustment knob

Historically, mercury barometers were used with a variety of standard gravities and temperature depending on the field of application – meteorology, physics etc. This led, in 1954, to the introduction of international standards, an example of which is BS2520.

The construction of a Fortin barometer is such that it reads pressure directly when the whole device is at 0 °C and at a gravity of 9.80665 ms^{-2} (standard conditions).

Correction tables for temperature allow for thermal expansion of the mercury ($\beta = 0.0001818$), and the brass scale ($\alpha = 0.0000184$).

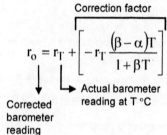

Correction factor

$$r_o = r_T + \left[-r_T \frac{(\beta - \alpha)T}{1 + \beta T} \right]$$

Corrected barometer reading

Actual barometer reading at T °C

Correction tables for gravity are expressed in terms of the latitude of the location of the instrument and the height above sea level. The acceleration due to gravity as a function of latitude ϕ^o is given by:

$$g_{\phi,0} = 9.80616\left(1 - 0.0026373\cos 2\phi + 0.0000059\cos^2 2\phi\right)$$

At a height Z m above sea level and at latitude ϕ, the acceleration due to gravity is:

$$g_{\phi,z} = g_{\phi,0} - \left(0.000003086Z\right)$$

These values of g lead to corrections to the barometer reading taken at a height Z metres above sea level and at a latitude ϕ.

$$r_{\phi,0} = r_{\phi,Z} + \left[-r_{\phi,Z} \frac{0.000003086Z}{g_{\phi,0}} \right]$$

$r_{\phi,0}$ what the barometer would read at sea level at latitude ϕ.

$r_{\phi,Z}$ actual barometer reading at latitude ϕ and height Z m.

The correction to be applied to the barometer reading for standard gravity is:

$$r_n = r_{\phi,0} + \left[r_{\phi,0}\left(\frac{g_{\phi,0}}{9.80665} - 1 \right) \right]$$

r_n what the barometer would read if located where g = 9.80665 m/s^2 and at sea level.

$r_{\phi,0}$ actual barometer reading at latitude ϕ and at sea level.

If the atmospheric pressure is required at some point above or below the cistern of the barometer (i.e. the barometer cannot be easily moved into the desired position) then a further correction is to be made for the hydrostatic pressure of the air column between the two heights.

e.g. For a point 1 m above the cistern, the correction is −0.1181 mb, 1 m below: +0.1181 mb

1.5.7 Pressure

Simple switch type
The oil **pressure switch** is screwed into a drilling from the outlet side of the oil pump. The oil pressure switch consists of a diaphragm which opens switch contacts if the oil pressure is sufficient to overcome the force of an opposing spring.

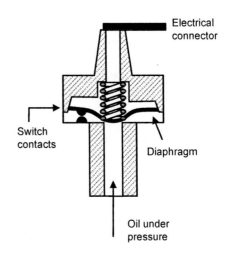

Bourdon gauge type
The **Bourdon gauge** consists of a tube bent into a coil or an arc. As the pressure in the tube increases, the coil unwinds. A pointer connected to the end of the tube can be attached to a lever and a pointer calibrated to indicate pressure.

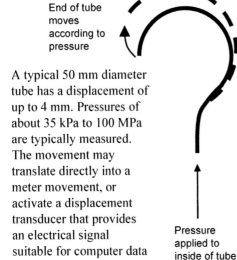

A typical 50 mm diameter tube has a displacement of up to 4 mm. Pressures of about 35 kPa to 100 MPa are typically measured. The movement may translate directly into a meter movement, or activate a displacement transducer that provides an electrical signal suitable for computer data acquisition.

The tube itself is made from brass and has a flattened elliptical or rectangular cross-section.

1.5.8 Industrial pressure measurement

Pressure is one of the most important process variables that need to be measured in industrial applications. The most common arrangement makes use of a diaphragm to which is bonded a **piezoresistive** displacement transducer. Bending of the diaphragm leads to an imbalance condition in a bridge circuit, the degree of which is a direct measurement of pressure.

The diaphragms are typically made from stainless steel which allows them to be used with water and other corrosive fluids. In some applications, **silicon diaphragm** sensors are available that are useful for high frequency measurements. These devices have their strain gauges bonded to them by atomic diffusion during manufacture.

The diaphragm can also be attached to a **piezoelectric crystal**. Pressure transducers of this type incorporate acceleration compensation which minimises the response of the device to **vibration**. They are useful for the measurement of pressure variations occurring under conditions of high static pressure.

When **mounting** a pressure transducer, it should be noted that any mechanical loading (other than the pressure being measured) will cause a deflection of the diaphragm resulting in an error in the signal. It is common practice to monitor the output of the device (at zero pressure) during mounting and tightening to ensure that no **mechanical stressing** of the housing and subsequently the diaphragm occurs.

Different types of pressure

Gauge:	The pressure measured relative to ambient atmospheric pressure.
Absolute:	Absolute pressure is equal to gauge pressure added to atmospheric pressure.
Differential:	The pressure measured relative to a reference pressure.
Proof:	The maximum pressure that may be applied for the device to remain within specifications.
Burst:	The maximum pressure that may be applied without physical damage to the transducer.

1.5.9 Sound

Microphones are pressure transducers designed for rapid changes in pressure at low amplitudes. For the professional **sound** industry, the frequency response and directional characteristics of the microphone are the most important parameters. Most microphones in use are of the **condenser** type.

Condenser

A metal diaphragm forms one plate of a **capacitor**, the other plate is fixed. Sound waves cause the diaphragm to move. If the diaphragm moves towards the fixed plate, then, if the voltage across the plates is a constant, this causes an increase in the field strength between the plates since $V = Ed$.

An increase in field strength draws more charge onto the plates thus resulting in a current flow in the connecting wires to the microphone. When the diaphragm moves away from the fixed plate, the current flow is reversed. The alternating current has frequency components equal to that of the incoming sound waves. For high frequency work, an AC carrier signal is applied across the plates. The sound waves are then represented by a **modulation** of the carrier.

Piezoelectric

A diaphragm is attached to a **quartz crystal**. Displacements of the crystal arising from sound waves generate output voltage proportional to the amplitude of sound waves. Some crystal microphones have a preamplifier in them to reduce noise pickup by leads to the main power amplifier.

Moving coil

A diaphragm is attached to a coil which moves relative to a fixed **permanent magnet**. The voltage induced in the coil is proportional to the amplitude of the sound wave.

Carbon button

Sound acts on a diaphragm which acts on an enclosed volume of **carbon granules**. Contact resistance between the granules depends upon the pressure. If a DC bias voltage is applied, the alternating resistance produces an AC signal which is proportional to the sound intensity.

1.5.10 Flow

The measurement of the flow of gases or liquids can be performed using a restriction which causes a **pressure drop**. The volume flow rate is usually proportional to the square root of the pressure difference. These types of transducers are called **differential pressure** or **dp** flowmeters. They may employ orifices, nozzles, pitot tubes and centrifugal elbows.

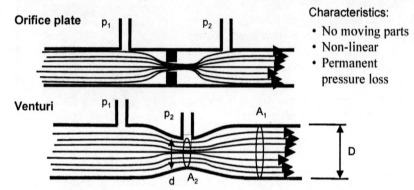

Orifice plate p_1 p_2

Characteristics:

• No moving parts
• Non-linear
• Permanent
  pressure loss

Venturi p_1

p_2 A_1

D

d A_2

For an incompressible fluid, and frictionless flow, the theoretical volume flow rate Q_{th} is:

$$Q_{th} = \frac{A_2}{\sqrt{1 - \left(\dfrac{A_2}{A_1}\right)^2}} \sqrt{\frac{2(P_1 - P_2)}{\rho}}$$

For the **orifice plate**, A_2 is the area of the **vena contracta**.

For real fluids, but still liquids, a correction factor C, being the **discharge coefficient**, is applied:

$$Q_{actual} = CQ_{th}$$

The discharge coefficient depends on the type of meter (orifice or venturi) and is usually measured experimentally.

In one type of commercially available device, a tube is placed perpendicular to the flow stream. Holes in each side of the tube face upstream (high pressure) and downstream (low pressure) leading to separate chambers inside the tube structure. The pressure differential is measured and calibrated to provide a flow rate. The cross-sectional shape of the tube is optimised to suit a wide range of fluid viscosities.

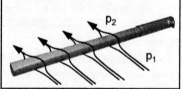

p_2

p_1

Positive displacement

Positive displacement flowmeters and rotating vane type flowmeters use the physical movement of a vane or piston as an indication of flow rate. In a typical device, two impellers are rotated by the flowing liquid. Magnets in the impellers activate an external sensor which generates a pulsed output signal.

Each pulse represents a known volume of liquid that is captured between the lobes of the impellers and the pulse count rate can thus be calibrated to provide flow rate in litres/minute. There should be enough back pressure on the outlet side of the flowmeter so that no gas pockets are formed during its operation. The positive displacement flowmeter is only suitable for liquid flow measurements.

Turbine

A **turbine flowmeter** contains a rotating vane whose angular velocity is measured and converted into flow rate. This type of flowmeter is applicable for both liquid and gas flow measurements and is suitable for very low flow rates.

Thermal

Mass flow rate can be determined by measuring the **temperature drop** of a heated sensor. The technique is suitable for measuring gas and liquid flow. Some sensors of this type are used as a flow/no flow switch that can be used to activate safety devices and level sensing. The technique uses no moving parts.

Ultrasonic

In one ultrasonic method, a beam is directed into the flowing fluid at an angle and the **doppler shift** in frequency of the reflected beam is an indication of flow rate. In another method, the time taken for an ultrasonic pulse to travel from a transmitter to a receiver downstream is used as a measure of flow.

Turndown is the ratio between the minimum and maximum flow conditions in a system. A 10:1 turndown ratio means that the maximum flow rate is 10 times the minimum flow rate in a system. A good flowmeter will be accurate for a turndown ratio of approximately 150:1.

Float

The flowing fluid acts against the mass of the float that is inserted in the stream. A calibrated scale shows the flow rate directly.

The flowmeter is mounted vertically with the inlet connection at the bottom of the unit.

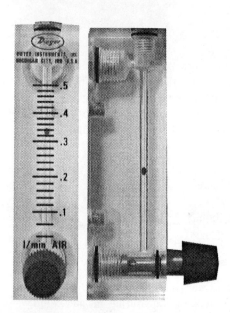

This type of flowmeter is very sensitive to the arrangement of inlet and discharge piping configurations. The inlet piping should be as large a diameter as the inlet to the meter and be as straight as possible – free from elbows, kinks and any other restrictions.

For gas flow measurements, the outlet or discharge pipe should be as large as possible to minimise back pressure. The gauge markings are calibrated for an outlet into standard atmospheric pressure. For liquid flow measurements, a moderate back pressure is permissible.

Bernoulli's equation

This equation governs the physics of streamline or laminar flow.

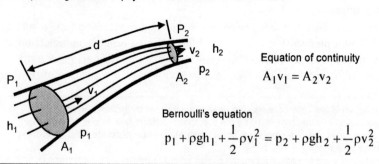

Equation of continuity

$$A_1 v_1 = A_2 v_2$$

Bernoulli's equation

$$p_1 + \rho g h_1 + \frac{1}{2}\rho v_1^2 = p_2 + \rho g h_2 + \frac{1}{2}\rho v_2^2$$

1.5.11 Level

The measurement of the **level** of liquids in tanks is a very important sensor and transducer application for process control. Various methods are available:
- Float
- Differential pressure
- Ultrasonic
- Capacitance
- Radar
- Ultrasonic

In a **float** system, the float typically acts upon a displacement transducer directly or is mounted on a float arm that in turn operates a displacement transducer. Float systems are mechanical devices that are prone to wear and corrosion. The displacement transducer can be of the resistive type that offers a continuously varying signal, or simple switches that indicate an on/off condition for upper level and lower level limits.

The **differential pressure level** transducer measures the pressure difference between an upper and lower position in a tank (or the atmosphere for a vented tank), and knowing the density of the fluid, the height of the fluid can be determined.

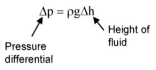

Ultrasonic **level transducers** determine level by measuring the length of time it takes for an ultrasonic pulse to be detected by a piezoelectric transducer after reflecting from the fluid surface. While there are no moving parts, vapours and turbulence affect the accuracy of the device.

In a **capacitance** type level meter, the wall of the tank is used as one plate of a capacitor, and an electrode placed in the centre of the tank as the other forming a capacitor. Changes in level of the liquid (which must be non-conducting) alter the capacitance (due to a change of permittivity of the insulating medium or dielectric) of a connecting circuit driven at RF frequencies. For conductive liquids, the probe is covered with an insulating sheath (which becomes the dielectric) and a change in level is registered as a change in capacitance due to a change in effective area between the probe and the grounded tank walls.

Radar systems work on a similar principle to the ultrasonic type except that electromagnetic waves are used to determine the time taken for a reflection from the liquid surface to be received.

1.5.12 Review questions

1. A strain gauge element has a gauge factor of 2.0 and an unstrained resistance of 150 Ω. If the change in resistance of the element at maximum strain is 5 Ω, determine the maximum strain that the element is designed to measure.   (Ans: 1.67%)

2. Calculate the required spring stiffness k and damping constant λ for an accelerometer that has a natural resonance at 10 Hz, a damping ratio of 0.8, and a seismic mass of 5 grams.   (Ans: 0.197 N/m, 0.05 Ns/m)

3. The casing of a compressor is vibrating sinusoidally with a displacement amplitude of 10^{-4} m and a frequency of 500 Hz. Calculate the amplitude of the acceleration.   (Ans: 100 g)

4. Would you expect the resonant frequency of a piezoresistive force transducer to be above or below that of a piezoelectric force transducer and why?

5. A bellows is used to create a force in a system without contributing significantly to the stiffness in the system. If the bellows can be considered a series connection of 5 springs, each of stiffness 1 Nm^{-1}, calculate the stiffness introduced into the system by the bellows.

6. The expression describes the sensitivity of a piezoelectric force transducer. Letting $s = j\omega$, determine an expression for the sensitivity as a function of ω and indicate the general features of this response on a freehand graph. Assume ω_o and ξ are constants.

$$\frac{\Delta V_{out}}{\Delta F} = \frac{1}{k} \frac{\omega_o^2}{s^2 + 2\xi\omega_o s + \omega_o^2}$$

7. A Fortin barometer is used to measure atmospheric pressure at latitude 53 °N and at a height 10 m above sea level. The vernier scale on the barometer reads 992.4 mbar and a nearby thermometer reads 19.8 °C. Calculate the corrected atmospheric pressure.   (Ans: 989.84 mbar)

8. How does flow rate vary with the pressure drop in a restriction type flow transducer (a semi-quantitative answer please, e.g. square, linear, exponential etc.).

9. If the discharge coefficient for a particular orifice plate of 1 cm diameter inside a 5 cm pipe is 0.78, calculate the differential pressure for a flow rate of water of 1 litre/sec.   (Ans: 8.34 kPa)

Part 2: Interfacing

2.0 Interfacing

It is common practice to use a computer to record measurements from a transducer. Transducers generally provide an analog signal that must be converted to digital format for data storage and analysis. The connection between the transducer and the computer is called the computer **interface**.

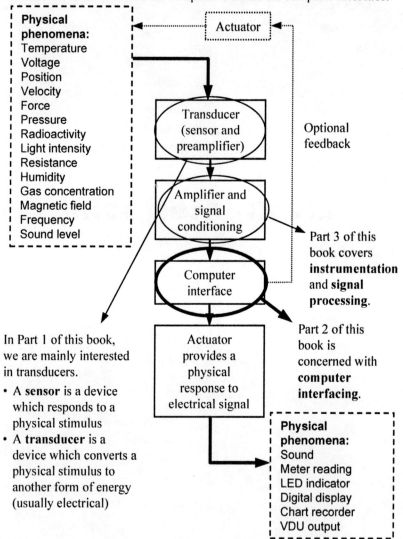

Physical phenomena:
Temperature
Voltage
Position
Velocity
Force
Pressure
Radioactivity
Light intensity
Resistance
Humidity
Gas concentration
Magnetic field
Frequency
Sound level

Actuator

Transducer (sensor and preamplifier)

Amplifier and signal conditioning

Computer interface

Actuator provides a physical response to electrical signal

Optional feedback

Part 3 of this book covers **instrumentation** and **signal processing**.

Part 2 of this book is concerned with **computer interfacing**.

Physical phenomena:
Sound
Meter reading
LED indicator
Digital display
Chart recorder
VDU output

In Part 1 of this book, we are mainly interested in transducers.

• A **sensor** is a device which responds to a physical stimulus
• A **transducer** is a device which converts a physical stimulus to another form of energy (usually electrical)

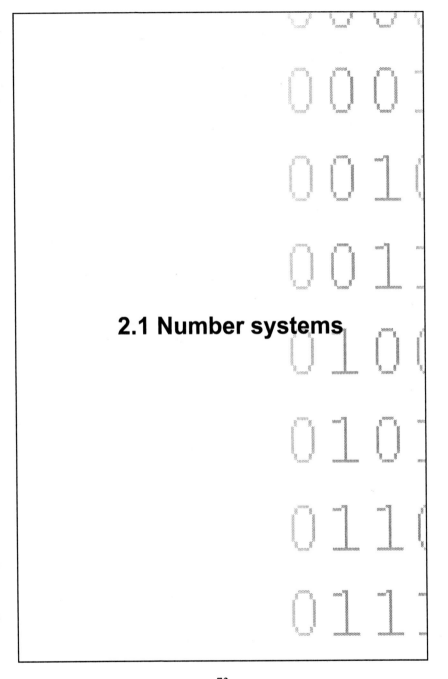

2.1 Number systems

2.1.1 Binary number system

There are ten digits in the decimal numbering system. In the **binary system**, there are only two, 0 and 1. Each digit in a binary number is called a **bit**. Computers consist of millions of transistor switches that can be either **on** or **off**, or **true** or **false**, and as a consequence they employ the two available digits in binary number system to represent the states of these switches.

Decimal	Binary
0	0000
1	0001
2	0010
3	0011
4	0100
5	0101
6	0110
7	0111
8	1000
9	1001
10	1010
11	1011
12	1100
13	1101
14	1110
15	1111

Bits can be arranged to provide a numerical code that can be used to convey information. A particularly popular code is the **ASCII** code used to represent decimal digits, alphabetic characters.

Since combinations of only two digits are used to represent binary numbers (i.e. 0 and 1), they tend to be rather cumbersome when large numbers are to be represented. For example, the number 26 in decimal is given by:

$$26_{10} = 11010$$

A group of 8 bits occurs very frequently in computer systems and is called a **byte**. Groups larger than 8 bits, such as 12 or 16 bits, are called **words**. The left most digit in binary representation is called the most significant bit, or **msb**, since it has the most weight in determining the magnitude of the number. The right most digit is called the least significant bit or **lsb**.

Powers of 2	
2^2	4
2^3	8
2^4	16
2^5	32
2^6	64
2^7	128
2^8	256
2^9	512
2^{10}	1024
2^{11}	2048
2^{12}	4096
2^{13}	8192
2^{14}	16 384
2^{15}	32 768
2^{16}	65 536

It is convenient to express large numbers arising from binary arithmetic by a convenient factor which happens to be $2^{10} = 1024$. For example, 65 536 divided by 1024 is 64 k where the 'k' means divided by 1024. For a 24 bit word, we divide 16 777 216 by 1024 twice to obtain: 16M where M means mega. 1024 bytes is one **kilobyte** and thus, 640 kb is really 640 × 1024 = 655 360 bytes. 1 **megabyte** (Mb) is 1 × 1024 × 1024 = 1 048 576 bytes. 64 kb = 65 536 bytes, which, if numbered sequentially, would be referenced from 0 to 65 535.

2.1.2 Decimal to binary conversion

The binary numbering system has only two digits, 0 and 1. It is easy to convert from **decimal to binary** by repeated divisions by 2. Start towards the right side of the page and work to the left.
Example: Convert 26_{10} to binary:

	26	Quotient
	÷2	
	13	Answer
	0	Remainder

Transfer **Answer** to next **Quotient** column to the left and divide by 2.

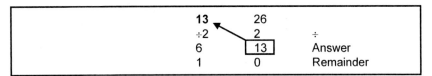

Repeat until zero obtained as **Answer**.

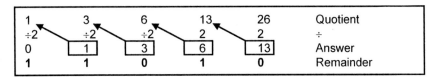

The result is given by the **Remainder**: $26_{10} = 11010$.

For **binary to decimal**, each binary position, starting at the least significant bit, represents a power to which the base 2 should be raised – starting from 0. Example: Convert 11010 to decimal:

$$= 1(2^4) + 1(2^3) + 0(2^2) + 1(2^1) + 0(2^0)$$
$$= 26$$

The numbering of bits in a binary number from 0 to 7 starting from the right and going to the left may seem a little backwards but ensures that each bit is raised to the power given by the bit position when converting to decimal.

2.1.3 Hexadecimal

Programming at assembly or machine language levels often entails
working with groups of 4, 8 or 16 bits at a time. For this reason, it is
simpler to use a numbering system which has as its base 4 bits, which is a
maximum of 16 combinations from 0000 to 1111. The **hexadecimal**
numbering system is based on 16 combinations of 4 bits and uses letters to
signify numbers greater than decimal 9. Thus, single digit numbers go from
decimal 0 to 9 but letters A through F are used to represent numbers greater
than 9. The term "hexadecimal" is a combination of **hex** meaning six and
decimal meaning, of course, ten.

Decimal	Hex	Binary
0	0	0000
1	1	0001
2	2	0010
3	3	0011
4	4	0100
5	5	0101
6	6	0110
7	7	0111
8	8	1000
9	9	1001
10	A	1010
11	B	1011
12	C	1100
13	D	1101
14	E	1110
15	F	1111

Note: We are only
concerned with integer
numbers here.
Methods of expressing
fractions need not
concern us for the
moment.

Hexadecimal (hex) numbers may be written with a leading $ sign to
distinguish them from decimal numbers. For example, $FF is decimal 255
or 1111 1111 in binary. Binary numbers may be specified with a leading
% and decimal integers with . which in this case is not to be taken as a
decimal point. Hex and binary numbers may also be represented by a
trailing "h" or "b" respectively.

2.1.4 Decimal to hex conversion

Convert from decimal to hex:

Repeated divisions by 16. Start at right of page and work towards left. Stop when 0 obtained as answer. The result is given by the Remainder.
Example: Convert the number 26 to hex:

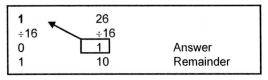

Answer: 1A (10 becomes A in hex).

Convert from hex to decimal:

Each binary position, starting at the least significant bit, represents a power to which the base 16 should be raised – starting from 0.
Example: Convert 1A to decimal:

$$1A = 1(16^1) + 10(16^0)$$
$$= 26$$

Convert from hex to binary:

Arrange each hex digit in groups of four and use the hex table.
Example: Convert 1A to binary:

```
   1     A
0001  1010
```

Answer: Obtain binary equivalents from hex table and write the binary in groups of four. This group of 8 bits is called a **byte**. The answer is 0001 1010 and the leading zeros can be discarded if desired. However, it is sometimes convenient to keep the binary digits grouped in lots of four and we write: 0001 1010. The groupings are for our convenience only.

Dec	Hex	Bin
0	0	0000
1	1	0001
2	2	0010
3	3	0011
4	4	0100
5	5	0101
6	6	0110
7	7	0111
8	8	1000
9	9	1001
10	A	1010
11	B	1011
12	C	1100
13	D	1101
14	E	1110
15	F	1111

2.1.5 2's complement

The **2's complement** is a special operation performed on a binary number which yields a new binary number, the significance of which will be explained shortly. The 2's complement is found by reversing all the bits in a binary number (called the **complement** or the 1's complement) and then adding 1 to the result. Example: What is the 2's complement of 13?

```
Original number 13    0  0  0  0  1  1  0  1
invert all bits       1  1  1  1  0  0  1  0  ◄──────┐
add 1                                      1         │
2's complement        1  1  1  1  0  0  1  1         │
                                                     │
```

It so happens that the 2's complement can be used to represent the negative of a positive integer thus allowing a computer to perform a subtraction with a digital adding circuit. Taking the 2's complement of a number twice returns to the original number.

The binary number in this row is called the "1's complement"

Note that in our discussion so far, we have always been working with integers. Indeed, that is the only type of number that computers can work with. However, we need to represent fractions and very large numbers as well in everyday computing applications. How is this done? Briefly, the computer divides up numbers into two parts called the **mantissa** and the **exponent**. The format is very similar to **scientific notation** used by scientists and engineers. The number 7×10^5 (sometimes written 7E5) is 700 000. In this example, 7 is the mantissa and 5 is the exponent.

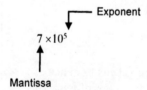

The computer stores the mantissa and the exponent in different places. In a simplified system, the mantissa may occupy 4 bytes of storage followed by a 1 byte exponent. Decoding of this format is typically done in software, although a specialised **maths co-processor** chip is fitted to most microcomputers to perform these conversion routines in hardware, and thus more quickly.

2.1.6 Signed numbers

An 8-bit memory location can cover the range of decimal integers from 0
to 255. To enable an 8-bit memory location to hold both positive and
negative numbers, the most significant bit (msb) is reserved and is called
the **sign bit**. A sign bit = 1 indicates a negative number. A sign bit = 0
indicates a positive number. The other 7-bit positions are used to represent
the magnitude of the number – but the way of doing this is different for
positive and negative numbers.

1. Positive numbers: The remaining 7-bit positions represent the magnitude of the number directly. 7 bits give a range 0 to +127.	**2. Negative numbers** are represented in 2's complement notation. Example: –40 is:

1. Positive numbers: The remaining 7-bit positions represent the magnitude of the number directly. 7 bits give a range 0 to +127.

```
0  0 1 0 1 0 0 0
```

0 msb indicates a positive number. The magnitude of the number is $2^3 + 2^4 = 40$.

2. Negative numbers are represented in 2's complement notation. Example: –40 is:

40 in binary notation	00101000
complement:	11010111
add 1	1
–40 in 2's complement:	11011000

The sign bit (msb) indicates a negative number.

Whether or not a particular binary bit pattern represents a signed or
unsigned number depends on the context in which it is being used. For
binary numbers starting with 0, there is no confusion since they have the
same value whether they are signed or unsigned numbers. For numbers
starting with 1, they may be interpreted as an **unsigned integer** or the 2's
complement representation of a negative number.

7F	0	1	1	1	1	1	1	1	+127
.	.	.	.	.	.	.	.	.	↑ .
.	.	.	.	.	.	.	.	.	.
.	0	0	0	0	0	0	0	1	2 1
0	0	0	0	0	0	0	0	0	0
FF	1	1	1	1	1	1	1	1	-1
.	1	1	1	1	1	1	1	0	-2
.	.	.	.	.	.	.	.	.	.
80	1	0	0	0	0	0	0	0	↓ -128

For signed number
representation, the
range that can be
covered by 8 bits is
–128 to +127.
Signed positive
binary numbers roll
over to represent
negative numbers
after +127.

The +1 step in finding the 2's complement takes into account ±0 possibilities
(i.e. 1's complement goes from –127 to –0, and +0 to +127).

2.1.7 Subtraction and multiplication

Subtraction

For a subtraction, the 2's complement is
added. Example: $43 - 40 = 43 + (-40)$

00101011	+43
11011000	−40 in 2's complement
100000011	answer = 11 (i.e. 3)

The extra bit on the left in the answer is called the **carry** bit. The carry bit is ignored in signed arithmetic but not in unsigned arithmetic.

Example: $40 - 43 = ?$

00101000	+40
11010101	−43 in 2's complement
11111101	add to find answer

Convert answer from above into decimal:

The **msb** is a sign bit which in this case indicates a negative number. To find out what this number is in decimal, we need to find the inverse 2's complement.

11111101	answer from above
00000010	complement
1	add 1
00000011	final answer is −3

Note that the final answer in decimal is −3 since the sign bit indicated that the number being stored was a negative number.

Multiplication

Multiplication and division by 2 in the binary number system is very easily done by a **shift**. Consider the product $2 \times 4 = 8$. Now, $4_{10} = 0100$ and shift to the left, gives 1000 (8_{10}). A shift to the right is a division by 2.

Multiplications with other numbers in binary is performed in exactly the same way as for decimal numbers e.g. $12 \times 6 = 8$.

This is analogous to positioning of decimal digits in the 1's, 10's and 100's columns.

							multiplicand 12_{10}
			1	1	0	0	multiplicand 12_{10}
			0	1	1	0	multiplier 6_{10}
			0	0	0	0	x by 0 with no shift
		1	1	0	0		x by 1 and shift left
	1	1	0	0			x by 1 and shift twice
0	0	0	0	0			x by 0 and shift thrice
1	0	0	1	0	0	0	+ for final result = 72

Multiplication involves repeated shifts left and additions. The process of division is very similar except that it involves repeated subtractions of the divisor.

Start by multiplying the multiplicand by the lsb of the multiplier. Repeat with other bit positions of multiplier and shift answers left one position each time.

2.1.8 Binary coded decimal (BCD)

Coding schemes are used to represent data in binary format. Although numbers may be expressed in the binary system directly, it is sometimes more convenient to use a coding scheme. A very well-known scheme for numerical data is the **binary coded decimal** system. The **BCD** code uses binary numbers 0 and 1 to represent decimal numbers 0 to 9. Each digit in a decimal number is transcribed into a 4-bit binary number.

Decimal	BCD
0	0000
1	0001
2	0010
3	0011
4	0100
5	0101
6	0110
7	0111
8	1000
9	1001

The main advantage of the BCD system is that the binary numbers in BCD are easily recognised and converted into decimal numbers because of their position.

The main disadvantage is that arithmetic operation on BCD encoded data is not so easily performed. BCD adders are required to perform arithmetic operations.

Note: Binary numbers above 1001 are not a part of the BCD system.

Consider the decimal number 2563. To represent this number in BCD is fairly straightforward. We simply write the binary numbers out in sequence for each digit in the decimal number.

$$2 \quad\quad 5 \quad\quad 6 \quad\quad 3$$
$$0010 \quad 0101 \quad 0110 \quad 0011$$

Each decimal number from 0 to 9 is represented by a four digit binary number. The weight, or contribution, of the msb in each binary number is $2^3 = 8$. The weight of the lsb is $2^0 = 1$. The weights of the other two bit positions are $2^2 = 4$ and $2^1 = 2$. The BCD code is sometimes referred to as an **8421 code** for this reason.

2.1.9 Gray code

The **Gray code** is well-known code originally used for encoding the angular position of a **rotary encoder**. Such an encoder may be constructed by a masked wheel whose concentric tracks are read by photo cells.

The main problem with the binary number system in this type of encoder is that there are many positions in which several tracks change their state at the same time. Thus, if a read operation occurs part way through a transition from one angular position to another, then the resulting error could be quite large.

Gray code

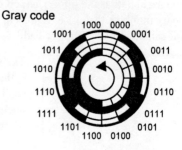

In the Gray code, only one track changes state at any one time during a rotation. Should a read error occur, then the resulting number will be in error by only one bit value. The Gray code is a **non-weighted** code. Any binary number can be converted into the Gray code, there is no upper limit to the number of code combinations. Conversion from Gray to binary can easily be done in a computer and so this code makes it ideal for this type of instrumentation purpose.

To convert from binary to Gray, we start at the msb and compare it to 0. If the msb is 0, then we write 0 as the msb for the Gray coded number, otherwise we write 1. We next compare the next msb and compare it to the msb. If they are equal we write a 0 in the position for the Gray coded number, otherwise, 1. We then compare each bit in the binary number to the bit just to the left of it and write 0 for a true comparison and 1 for a false. This procedure continues until the lsb is compared with the second bit.

Dec	Hex	Bin	Gray
0	0	0000	0000
1	1	0001	0001
2	2	0010	0011
3	3	0011	0010
4	4	0100	0110
5	5	0101	0111
6	6	0110	0101
7	7	0111	0100
8	8	1000	1100
9	9	1001	1101
10	A	1010	1111
11	B	1011	1110
12	C	1100	1010
13	D	1101	1011
14	E	1110	1001
15	F	1111	1000

2.1.10 ASCII code

The ASCII code is almost universally used to represent both numeric, character and special symbol data. The code is, in its standard form, a 7-bit code. **7 bits** gives 128 different combinations. The 8th bit is sometimes used as a **parity bit** for error detection. In the **extended ASCII** character set, the 8th bit (or msb) is used to create another 128 characters that contain mathematical and other special symbols.

7-bit ASCII code   msb

	0	1	2	3	4	5	6	7
0	NUL	DLE	Space	0	@	P	'	p
1	SOH	DC1	!	1	A	Q	a	q
2	STX	DC2	"	2	B	R	b	r
3	ETX	DC3	#	3	C	S	c	s
4	EOT	DC4	$	4	D	T	d	t
5	ENQ	NAK	%	5	E	U	e	u
6	ACK	SYN	&	6	F	V	f	v
7	BEL	ETB	'	7	G	W	g	w
8	BS	CAN	(	8	H	X	h	x
9	HT	EM	)	9	I	Y	I	y
A	LF	SUB	*	:	J	Z	j	z
B	VT	ESC	+	;	K	[	k	{
C	FF	FS	'	<	L	\	l	|
D	CR	GS	-	=	M	]	m	}
E	SO	RS	.	>	N	^	n	~
F	SI	US	/	?	O	_	o	DEL

lsb   Example: The number 4F, or 100
             1111, is the letter 'O'.

The first 32 characters in the code are control codes. These codes are interpreted by the device to which the data is being sent. For example, a printer receiving a CR code would execute a carriage return. Many microcomputer applications store their data in ASCII format. It is probably the most universally used method of representing numeric and character data for both storage and transmission purposes.

2.1.11 Boolean algebra

Digital electronic circuits contain components which act like high speed switches that process voltage levels **TTL** high (5 V) and TTL low (0 V). These circuits are thus suitable for representing the binary numbers 0 and 1. TTL high and TTL low may also represent **logic states** true and false and thus allow binary data to be processed using **Boolean algebra** in a digital circuit. The components of a digital circuit are called **logic gates**. Boolean algebra are laws which specify the interaction between logical states true (1) and false (0). **Truth tables** provide the rules for the Boolean operators.

Binary system:	
True	False
High	Low
Mark	Space
On	Off
0	1
5 V	0 V

A	B	A AND B	A•B
0	0	0	Output true
0	1	0	only if both A and B
1	0	0	are true
1	1	1	**AND**

A	B	A OR B	A+B
0	0	0	Output true
0	1	1	if either A or B are
1	0	1	true
1	1	1	**OR**

AND gate

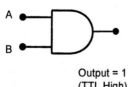

Output = 1
(TTL High)
when A and B
are both 1

A	B	A NAND B	
0	0	1	Output true if
0	1	1	both A and B are not
1	0	1	true
1	1	0	**NAND**

A	B	A NOR B	
0	0	1	True if A and B
0	1	0	are both not true.
1	0	0	
1	1	0	**NOR**

NOR gate

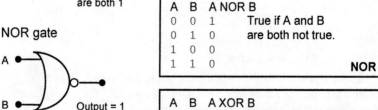

Output = 1
(TTL High)
when neither A
nor B are 1

A	B	A XOR B	
0	0	0	True if either A or B
0	1	1	are true but not both
1	0	1	together.
1	1	0	**XOR**

2.1.12 Digital logic circuits

Boolean algebra can be implemented using digital electronic circuits using combinations of **logic gates**.

e.g. A combination of NAND gates gives a logical **XOR** function.

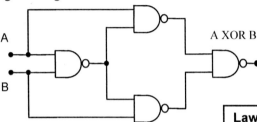

A   A XOR B

B

Truth table

A	B	O
0	0	0
0	1	1
1	0	1
1	1	0

In the circuit below, the XOR function is used to add binary digits A and B. The AND gate indicates whether or not there is a **carry** bit. This circuit is a **half adder**.

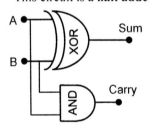

A

B

Sum

Carry

Truth table

A	B	S	C
0	0	0	0
0	1	1	0
1	0	1	0
1	1	0	1

Laws of Boolean algebra

$A + B = B + A$
$B \bullet A = A \bullet B$

$(A + B) + C = A + (B + C)$
$(A \bullet B) \bullet C = A \bullet (B \bullet C)$

$A + AB = A \bullet (1 + B) = A$
$A \bullet (A + B) = A$

$A \bullet (B + C) = A \bullet B + A \bullet C$
$A + (B \bullet C) = (A + B) \bullet (A + C)$

$A + A = A$
$A \bullet A = A$

$A \bullet \overline{A} = 0$
$A + \overline{A} = 1$
$\overline{\overline{A}} = A$

$0 + A = A$
$1 \bullet A = A$
$1 + A = 1$
$0 \bullet A = 0$

$A + \overline{A} \bullet B = A + B$
$A \bullet (\overline{A} + B) = A \bullet B$

De Morgan's theorem

$\overline{(A + B)} = \overline{A} \bullet \overline{B}$
$\overline{(A \bullet B)} = \overline{A} + \overline{B}$

2.1.13 Review questions

1. How many numbers may be represented by a sequence of 8 binary digits?

2. Shown below are some decimal numbers. Fill in the columns assuming 2's complement notation.

Decimal	Binary	Hex
-1	.	.
127	.	.
28	.	.

3. Fill in the table below (assume 2's complement notation).

Binary	Decimal	Hex
1010	.	.
.	.	80
.	.	FF

4. Find the two's complement of $0E and show that by finding the 2's complement twice the original number is returned.

5. Consider the bit pattern 1011 1101. Determine another bit pattern (called a mask) which, when logically combined (using a Boolean expression) with the first, toggles the second most significant bit (from 0 to 1 or 1 to 0) but leaves the others unchanged.

6. Discuss the relative differences of the Gray code, the BCD code, and the ASCII code.

7. Design a logic circuit which implements the XOR function but using OR and NOR gates only.

8. Draw up the simplest logic circuit satisfying the truth table given:

A	B	C
0	0	1
0	1	0
1	0	1
1	1	1

2.1.14 Activities

1. Start the Microsoft Windows Calculator and set to scientific mode. Then select BIN for binary mode and WORD.

   Add the two 16-bit numbers below by first adding manually on paper and then using the calculator:

   ```
   1 0 1 1 1 0 1 1 1 0 1 0 1 0 1 0
   1 1 0 1 1 1 0 1 1 1 0 0 1 1 0 0
   ```

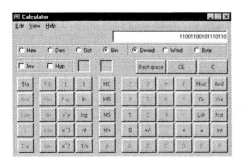

2. Set the Calculator to HEX mode and set to BYTE and then type in FFFF. Convert this number to Decimal and then Binary by selecting the appropriate buttons:

   Dec: _____

   Bin: _____

3. Perform a decimal subtraction using the Calculator which gives a negative result (say 8-10). Convert the answer to Hex and then to binary. What is the significance of these answers?

   Dec:
   Hex:
   Bin:

4. Consider the product 2 × 4 = 8. Verify that this multiplication, when performed in Binary mode using the Calculator, is the same as the one shift to the left of the binary representation of the decimal number 4 (you may need to find out how to use the "shift left" function of the Calculator using the Help topics).

5. Consider the bit pattern 1011 1101. Determine another bit pattern (a mask) which, when logically combined (using a Boolean expression) with the first, toggles the second most significant bit (from 0 to 1 or 1 to 0) but leaves the others unchanged. Use the Calculator to test this (using the Boolean operator keys).

2.2 Computer architecture

2.2.1 Computer architecture

The combination of functional components in a computer is referred to as the **architecture** of the microcomputer. The main functional components in a microcomputer are the **CPU** or central processing unit, **memory**, and input/output (**I/O**) devices:

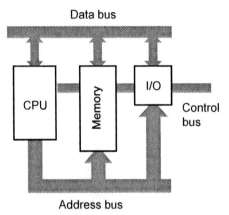

Data bus

Address bus

Each memory cell is capable of holding 8 bits, or 1 byte, of data. Memory cells are labelled with a unique **address**.

The **data bus** in the 8086 CPU is 16 bits wide and is **bi-directional**. It can transfer data in both byte and word length to and from the CPU and memory.

Binary information is transferred as TTL logic across wires called the **bus** – the **address** bus, the **data** bus and the **control** bus. When an address is placed on the address bus, the byte of information at or for that memory location is placed on the data bus. Signals on the control bus tell the CPU whether a read or write operation to that memory cell is required.

The address bus is **unidirectional** in that data is placed on it only by the CPU. The 8086 chip has a 20-bit address bus but the internal registers of the CPU are only 16 bits. A special **segmented memory** addressing scheme is used to obtain access to the full 1 MB memory.

The **control bus** carries various synchronisation and control signals, the only one of interest to us being the **read/write** signal. This is designated R/W (R/not write). During a read cycle, the processor receives data from either memory or a memory-mapped peripheral device. During a write cycle, the processor sends data to either a memory cell or a **memory-mapped** device.

The data bus is nominally 16 bits, but in 8088 machines, it is physically only 8 bits wide. Internally, the CPU transfers 2 bytes in sequence and operates as a 16-bit device. To further complicate matters, the data bus and the first 8 lines of the address bus on the 8088 are **multiplexed** (the same wires are used for both functions - but at different times). Later processors employ data buses up to 64 bits wide.

2.2.2 Memory

Each memory cell is capable of holding 8 bits, or 1 byte, of data. Memory cells are labelled with a unique **address**. When the microprocessor wishes to read or write data to a particular memory cell, it places the address of the required cell on the address bus, and the data to or for that memory location appears on the data bus. Internal circuitry ensures that only the memory cell whose address appears on the address bus receives or sends the data from or to the data bus.

The amount of memory that can be addressed by the CPU depends on the width of the address bus. The Intel 8086 CPU has a 20 bit address bus and is able to address a total of 2^{20} = 1 048 576 bytes (**1MB of RAM**). Each memory cell is numbered $00000 to $FFFFF. However, the internal working **registers** in the 8088 are only 16 bits wide and a special **segmented** memory addressing scheme is used to fit the 20-bit address data into the 16-bit registers. In contrast, the 80286 has a 24-bit address bus allowing 16 MB adressable RAM. Later processors employ a 32-bit address bus (486) giving 4 GB addressable memory while Pentium Pro and later processors have 64 GB addressable memory.

Address	data
00F101	32
00F100	D3
00E111	32

Program statements are always stored in the code segment. Data for programs is stored in the data segment etc.

Memory is divided into a number of **segments**, each of which is 64 kb in size:

- Code segment
- Data segment
- Stack segment
- Extra segment

Each segment is 64 k. Since the total address space is 1 048 576 bytes, there are a possible 1 048 576 /65 536 = 16 segments.

High memory

4-bit segment identifies one of 16 possible segments of memory.

```
0 1 0 0  . . . .
0 0 1 1  0000  0000 0000 0000
0 0 1 0  0000  0000 0000 0000
0 0 0 1  0000  0000 0000 0000
0 0 0 0  0000  0000 0000 0000
```

Thus, to specify a particular location in memory, all we need is a 4-bit segment **base address** and a 16-bit **offset**.

16 bit **Offset** identifies each of the 65536 individual memory locations in each segment.

Low memory

Unfortunately, it is not that simple. Segments need not start on the 64k boundaries shown above. Indeed, segments can start anywhere on a 16-bit boundary since the full 16-bit width of the CPU internal registers can then be utilised.

2.2.3 Segmented memory

Here's how segmented memory really works:

Segment: 1010 0111 1010 0100

Offset: 1000 1001 1100 1110

1. The 16-bit segment is multiplied by 16 to form a 20-bit **segment base address** by shifting to the left four times.

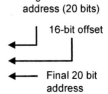

Segment base address (20 bits)

16-bit offset

```
  1010 0111 1010 0100 0000
+ ____ 1000 1001 1100 1110
  1011 0000 0100 0000 1110
```

Final 20 bit address

2. The 16-bit offset is added to the segment base address to obtain the 20-bit absolute address.

3. The **segmented address** is written with the segment followed by a colon ":" and then the offset. For example:

FFE2:01D0

Segment   Offset

The number obtained after the segment has been added to the offset is called the **absolute address**.

Note: Since segments can be specified with any 16-bit number, it is possible to have two different segmented addresses which refer to the exact same physical memory location! For example, 0010:0000 is the same as 0000:0100 which is memory location 100H.

Why is it useful to have segments start at any 16-bit address boundary? It permits a more efficient use of memory. For example, a particular program may not need a full 64k code segment. Some of this available memory may be used as the data segment by allowing the segments to overlap. By allowing a 16-bit number to specify the segment (instead of a 4-bit number), the start of each segment can be controlled to within a 16-bit boundary (instead of a 64 k boundary).

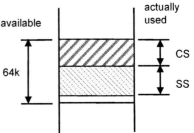

This example shows how the Stack Segment can be accommodated within a 64K block which is only partially used by the Code Segment.

2.2.4 Memory data

Each memory cell can store 8 bits, or 1 byte, of data. The width of the data bus indicates how much data can be transferred during each memory **read/write operation**. The 8088 CPU has an 8-bit data bus but can actually process 16 bits at a time using its 16-bit internal registers. The 80286 has a full 16-bit data bus. The 80486 has a 32-bit data bus and Pentium processors have a 64-bit data bus.

The contents of memory are interpreted by the CPU as either:

Data	Logical (1s and 0s indicating T and F) or Numeric – signed or unsigned integers as binary numbers.
Instructions	An assembly or machine language opcode.
Address	A "pointer" to which the CPU goes to get the data required.

Groups of bits larger than a byte are called **words**. In a 16-bit machine, the term word is used to describe 16-bit (2-byte) data and the term **long word** or **double words** for 32-bit (or 4-byte) data. In the 8086 architecture, words are stored in memory with the higher byte in the higher numbered address.

Address	data
06001	3E
06000	01

In the example here, referencing a word at 06000H, one would obtain: 3E01H. When words and double words start at an address that is a multiple of 4, they are **aligned**.

Double words take up four memory locations and are stored with the higher word at the higher address pair. Within each word, the higher byte is stored at the higher numbered address. It is usual to write memory addresses from the bottom upwards.

68000

The width of the address bus on the MC 68000 chip is effectively 24 bits (3 bytes). (On the 68000 chip, the address bus is only 23 bits wide numbered A1 to A23 with an internal A0 bit which controls the way the 16-bit data bus is used. The effective bus width is thus 24 bits.) The number of addressable memory locations is:

No. addresses $= 2^{24}$
$= 16\ 777\ 216$

which are numbered from 0 to 16 777 125 or in hex $000000 to $FFFFFF. Each address refers to a data space of one byte (or 8 bits) giving 16 MB of RAM.

2.2.5 Buffers

Binary signals are transmitted between the CPU and memory cells across the address and data buses. Proper communication of data requires that one and only one memory cell, the one whose address is present on the address bus, has direct connection to the data bus at any one time. **Decoding circuitry** determines the location of the desired memory cell to be activated. Activation of a single memory cell entails connecting the cell to the data bus and ensuring that all other cells are effectively disconnected. This connection procedure is carried out by **tri-state buffers**.

The table below shows that the connection between the data bus and a memory location must be set at a high impedance when the **chip-select** signal CS is low and pass through the data when chip-select is high.

Data	CS	Connection
0	0	High impedance
1	0	High impedance
0	1	0
1	1	1

Buffers isolate memory cells from the data bus and also allow data to pass through during **read/write** operations. Since the data bus must pass data in both directions, its connections to the bus and the memory cell must be capable of being at TTL high and TTL low (to represent logic levels 0 and 1). When a memory location is not selected (by the decoders), the buffer must effectively disconnect the memory cell from the data bus by inserting a high impedance. The term **tri-state** means that the connection made by the buffer can be either TTL high, low or high impedance.

A simple example of tri-state logic can be made using two transistors as shown:

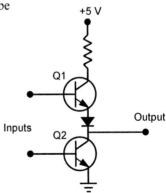

Q1	Q2	Output
Off	Off	High impedance
Off	On	0 V
On	Off	+5 V

2.2.6 Latches

A **latch** is a device which holds the data that appears on its input terminals. A memory cell in the microcomputer system is a latch. Typically, signals destined for storage in memory cells appear on the data bus momentarily and then disappear. The timing of the signals is regulated by the internal **clock** which runs at speeds typically in the MHz range. The decoding circuitry determines which buffer is to be activated. The activated buffer in turn connects the latch input terminals to the data bus. The signals on the data bus are transferred through buffers to the latch circuit which stores the signals on its output terminals.

A latch circuit can be implemented using a series of RS **flip-flops**. In this figure, the 4 bit data at D_3 to D_0 is transferred to Q on the clock pulse. When a bit D is logic 1, $S = 1$ and $R = 0$ and the output Q becomes 1. When D is logic 0, $S = 0$ and $R = 1$ and the output $D = 0$.

An **octal latch** has 8 inputs and 8 outputs. The data latch enable (DLE) pin, when set high, copies the voltage levels on the input pins to the corresponding output. The latch circuitry retains the signals on the output pins even if the input signals disappear and DLE goes low. It is important that DLE is set when data appears on the input. DLE is typically timed to go high when data appears on the data bus. The clock signals are used to synchronise this timing.

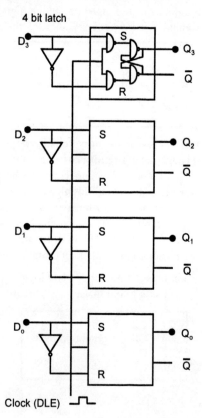

4 bit latch

Clock (DLE)

2.2.7 Flip-flop

Flip-flops can be used to represent binary numbers. An RS flip-flop is a digital circuit which is stable in one of two states – set or **reset**. Such a circuit can be made using NAND gates. A **truth table** summarises the action of flip-flop. The voltage of one of the outputs can be used to represent or store a binary digit since it can be either voltage high (logic 1) or low (logic 0) and will remain at that setting until signals on the input, which only last for a short time, set or reset the outputs.

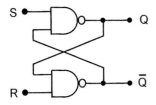

Action table (RS):

R	S	
0	0	not used
0	1	$Q = 0; \bar{Q} = 1$
1	0	$Q = 1; \bar{Q} = 0$
1	1	no change

A microcomputer uses a **clocked flip-flop** to synchronise the action of the circuit.

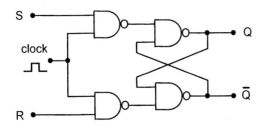

S

clock

R

Data at terminal S gets transferred to Q on the clock pulse and remains at Q even if the signal at S disappears and the clock goes low.

Action table (clocked RS):

R	S	
0	0	no change
0	1	$Q = 1; \bar{Q} = 0$
1	0	$Q = 0; \bar{Q} = 1$
1	1	not used

Note: This action table is different to the ordinary NAND flip-flop. Here, R = S = 1 is the "not used" state.

The data stays at Q because when the clock pulse goes low, the flip-flop circuits *within the chip* are at S = R = 1 (due to the NAND gates on the clock stage). Only when the clock goes high do the flip-flops react to the logic signals at D on the latch.

2.2.8 Input/Output (I/O)

There are three methods of handling input/output devices. The first is by
the use of **ports**. In an 8088-based microcomputer, ports are identified
using a 16-bit **port number**. Thus, there are a total of 65 536 available
ports numbered 0 to 65 535. The CPU uses a signal on the control bus to
specify that the information on the address bus and data bus refers to a port
and not a regular memory location. The port with the specified number
then receives or transmits the data from its own inbuilt memory which is
not part of the main computer's memory. However, some devices use
main memory for their own use and thus data for these devices may be
specified using regular memory addresses. This is the second method used
for I/O. Such devices (e.g. video adaptors) are called **memory-mapped**
I/O devices.

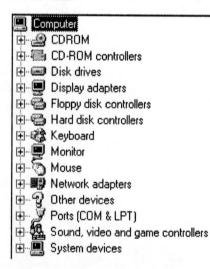

This list shows the range of input
and output devices for a typical
desktop microcomputer. These
devices may often by memory-
mapped and data is written
from/to them via interface
adaptors. Some ports are not
memory mapped and a signal on
the control bus identifies to the
CPU that the address is that of a
port and not a memory location.

The third method of I/O involves
bypassing the CPU and writing or
reading directly from memory. This is
called direct memory access or **DMA**.
A special DMA controller IC is used to
regulate traffic on the bus for this
activity. A computer's disk drive
usually transfers data to or from
memory using DMA.

ISA and PCI bus

Part of the success of the original
IBM PC and the Intel
microprocessor family was due to
the use of open architecture,
made possible by the expansion
bus. The original **ISA** (Industry
Standard Architecture) provided a
16-bit bus. The introduction of the
Pentium CPU also saw the
introduction of the **PCI** (Peripheral
Component Interface) bus. This
bus supports 32-bit and 64-bit
data transfers with an increase in
data transfer rate over the ISA
bus.

2.2.9 Microprocessor unit (MPU/CPU)

The **central processing unit** organises and orchestrates all activities inside the microcomputer. Each operation within the CPU is actually a very simple task involving the interaction of binary numbers and Boolean algebra. A large number of these simple tasks combine to form a particular function which may appear to be alarmingly complex.

The "MPU" is essentially the same thing as the more familiar and general term "CPU" (CPU applies to *any* computer, and not just a *micro*computer).

The **CPU** is responsible for initiating transferring data to and from memory and input/output devices, performing arithmetic and logical operations, and controlling the sequencing of all activities. Inside the CPU are various subcomponents such as the **arithmetic logic unit (ALU)**, the **instruction decoder**, internal **registers** and various control circuits which synchronise the timing of various signals on the buses.

```
CPU

Instruction decoder      Address registers
Arithmetic logic unit    Pointers
Registers                Flags
                         Instruction pointer
```

80X86 CPU development

1972	Intel introduces the 4004 with a 4-bit data bus, 10,000 transistors.
1974	8080 CPU has 8-bit data bus and 64 kb addressable memory (RAM).
1978	8086 with a 16-bit data bus and 1 MB addressable memory, 4 MHz clock.
1979	8088 with 8 bit external data bus, 16-bit internal bus.
1982	80286, 24 bit address bus, 16 MB addressable memory, 6 MHz clock.
1985	80386DX with 32-bit data bus, 10 MIPS, 33 MHz clock, 275×10^3 transistors
1989	80486DX 32-bit data bus, internal maths coprocessor, $>1 \times 10^6$ transistors, 30 MIPS, 100 MHz clock, 4 GB addressable memory.
1993	Pentium, 64-bit PCI data bus, 32-bit address bus, superscalar architecture allows more than one instruction to execute in a single clock cycle, hard-wired floating point, $>3 \times 10^6$ transistors, 100 MIPS, >200 MHz clock, 4 GB addressable memory.
1995	Pentium Pro, 64-bit system bus, 5.5×10^6 transistors, dynamic execution uses a speculative data flow analysis method to determine which instructions are ready for execution, 64 GB addressable memory.
1997	Pentium II, 7.5×10^6 transistors with MMX technology for video applications 64 GB addressable memory.
1999	Pentium III, 9.5×10^6 transistors, 600 MHz to 1 GHz clock.
2000	Pentium 4, 42×10^6 transistors, 1.5 GHz clock.
2001	Xeon, Celeron processors, 1.2 GHz, 55×10^6 transistors.

2.2.10 Registers

The 8088/6 CPU has 14 internal registers. All the registers are 16 bits.
Registers are used to hold data temporarily while the CPU performs
arithmetic and logical operations.

Data registers

	15		0
AX	AH	AL	
BX	BH	BL	
CX	CH	CL	
DX	DH	DL	

The data registers may be divided into two 8-bit registers depending on
whether the CPU is working with 8-bit, or 16-bit data. Within the 16-bit X
registers, the 8-bit registers are AL, BL, CL and DL, and AH, BH, CH, and DH.
Each half of the X registers may be separately addressed using L and H labels.

- AX is the **accumulator** and is used as a temporary storage space for
  data involved in arithmetic and string (character) operations.
- BX is the **base** register and is often used to hold the offset part of a
  segmented address during memory transfer operations.
- CX is the **count** register and is used as a counter for loop operations.
- DX is the **data** register and is a general purpose 16-bit storage
  location used in arithmetic and string operations.

Segment registers

	15	0
CS		
DS		
SS		
ES		

- CS is the **code segment** and contains the base address of the segment of
  memory that holds the machine language program that is being executed.
- DS is the **data segment** and contains the base address of the segment of
  memory where current data (such as program variables) are stored.
- SS is the **stack segment** and contains the base address of the stack
  segment which is used to hold return addresses and register contents
  during the execution of subroutines within the main program.
- ES is the **extra segment** and contains the base address of the extra
  segment which is used to supplement the functions of the data segment.

Offset registers

	15		0
IP		— — — — — — — — — — — — — — — —	
SP		— — — — — — — — — — — — — — — —	
BP		— — — — — — — — — — — — — — — —	
SI		— — — — — — — — — — — — — — — —	
DI		— — — — — — — — — — — — — — — —	

- IP is the **instruction pointer** (or program counter) which provides the offset address into the code segment (CS) for the address of the next instruction to be executed in a machine language program.
- SP is the **stack pointer** and together with the base pointer (BP) provide the offset into the stack segment (SS). This is the current location of the top of the stack.
- BP, the **base pointer**, is used in conjunction with the stack pointer to provide an offset into the stack segment.
- SI and DI are **index** registers and are used (usually in conjunction with a data register) to provide an offset into the data segment for the processing of long string characters.

Flags are individual bits which are used to report the results of various comparisons and processes done by the CPU. Program statements may then branch depending on the status of these flags. Although the flags themselves are individual bits, they are arranged together in the form of a 16-bit register so that their contents may be easily saved and restored whenever necessary (e.g. while a subroutine is being executed).

Flags

			OF	DF	IF	TF	SF	ZF		AF		PF		CF
__	__	__							__		__		__	

Carry flag	arithmetic carry out
Parity flag	even number of 1s
Auxiliary carry flag	used for BCD arithmetic operations
Zero flag	zero result or equal comparison
Sign flag	negative result or not equal comparison
Trap flag	generates single-step operation
Interrupt enable fag	interrupts enabled
Direction flag	decrement/increment index registers
Overflow flag	arithmetic overflow

When the status of flags is reported in diagnostic programs, a special notation is used. After executing an instruction, flags are either set (logic 1) or reset (logic 0). Instructions with the CPU's instruction set use these flags to jump to another section of the current program.

Flag			Set	Reset
CF	Carry		CY	NC
PF	Parity		PE	PO
AF	Auxiliary	AC	NA	
ZF	Zero		ZR	NZ
SF	Sign		NG	PL
IF	Interrupt		EI	DI
DF	Direction	DN	UP	
OF	Overflow	OV	NV	

Other instructions exist which allow a program to set or reset some of the flags. The flag register as a whole is usually pushed onto the stack when a subroutine executes and is then popped off the stack when the main program resumes.

The **stack** is a block of memory used for temporary storage. Saving data to the stack is called **pushing** and retrieving the data is called **popping**. Data pushed onto the stack can be popped off the stack on a last-in, first-out (**LIFO**) basis. The offset which represents the top of the stack is held in a register – called a **stack pointer**. The base address is the contents of the stack segment register. Thus, the segmented address of the top of the stack is given by SS:SP. The stack fills from high memory to low. The bottom of the stack is thus: SS:FFFF.

6000	A8
5FFF	34
5FFE	FE ←

In the example here, the data at offset address 5FFE is the top of the stack. The segment base address for this offset is the contents of the stack segment register.

Top of stack

When a **subroutine** is called within a program, the contents of CS and IP are pushed onto the stack. After the subroutine has finished, CS:IP are popped off the stack and thus execution of the main program resumes at the statement following the call to the subroutine. The subroutine itself can also save the contents of other registers by pushing them onto the stack and then popping them back before handing control back to the main program.

When data is pushed or popped from the stack, the stack pointer (SP) decrements or increments either by 2 or 4 depending on whether a word or a double word is being pushed or popped.

2.2.11 ROM

Conventional memory is called "**random access memory**" or RAM and is able to be read from or written to. ROM is **read only memory** and can only be read from. Data in ROM is burned in during manufacture of the memory chip. In an 8086 based microcomputer, there are a number of programs placed in ROM which allow the computer to do certain basic operations. For example, ROM typically contains:

* Start-up routines.
* BIOS (basic input/output services).
* BIOS extensions for additional equipment
  connected to the computer.

1. Start-up routines

When power is applied to the microcomputer, the first program to run is a power-on-self-test which does a memory check, initialises all support chips and the vector interrupt table, and finally loads the operating system.

2. BIOS

BIOS standards for **Basic Input/Output Services** and these services are a set of programs which allow application programs to interface with input/output devices connected to the computer in a consistent manner. BIOS programs are usually stored in ROM firmware. The operating system calls upon ROM BIOS routines using interrupts. The ROM BIOS relieves the application program of interfacing directly with memory locations to manage keyboard entry, video output, serial and parallel communications etc.

In this book, we are particularly interested in serial port communications. BIOS routines available for the serial port are:

Service
0   initialise serial port
    parameters
1   transmit character
2   receive character
3   get serial port status

For example, the serial port parameters (baud rate, parity, stop-bit, data bits) are specified in a bit pattern for a single byte which is placed in the AL register. When the service is called (using an interrupt), the initialisation information is read from AL and the BIOS programs the UART.

2.2.12 Interrupts

Servicing of I/O devices is usually done using **interrupts**. When an interrupt signal is received, the CPU suspends its activities and runs an **interrupt service routine** or **interrupt handler**. After the interrupt service routine has finished executing, normal execution is resumed.

There are three types of interrupts:

1. Microprocessor interrupts

These interrupts are initiated by various error conditions (such as a division by zero or arithmetic overflow). These interrupts are also known as **processor exceptions**.

2. Hardware interrupts

These interrupts are physically wired into the microcomputer. A special NMI interrupt has the highest priority and cannot be masked out by other interrupts. It is processed during critical hardware events such as a power loss.

3. Software interrupts

These interrupts are initiated by software to perform various operations such as writing to a disk file, reading from the serial port etc. These built-in interrupt handling routines are a part of the computer's **BIOS** – Basic Input/Output Services.

Interrupts are handled on a **priority** basis. The interrupt number determines its priority. High level interrupts cannot themselves be executed if a lower level or high priority interrupt is being processed.

The management of hardware interrupts is handled by a **programmable interrupt controller** chip: the 8259. This chip can be programmed to implement a variety of priority schemes and to accept level or edge-triggered interrupt signals. It determines which interrupt requires servicing and signals the CPU via the INTR line that an interrupt is pending. When an acknowledgement is received from the CPU, the 8259 places the interrupt number on the data bus and the CPU determines the address of the appropriate interrupt handler and the required interrupt service routine is then executed.

The 8259 controller can handle eight hardware devices. 8086-based microcomputers have one 8259 controller. 80286+ computers have two, with the second controller cascaded to interrupt channel 2 of the first giving access to 15 hardware devices.

An **interrupt vector table** contains the address pointer for the interrupt service routines associated with each of the 256 available interrupts. The interrupt vector table is usually located in low memory. Interrupt vectors 0 to 31 are usually reserved for microprocessor interrupts. The remainder can be used for hardware or software interrupts.

The interrupt **type number** determines its place within the interrupt vector table and its **priority** (with the exception of the NMI interrupt (2), but has the highest priority due to its direct connection with the CPU).

Microprocessor interrupts are divided into **fault, trap** or **abort** conditions. For fault conditions, the instruction that caused the fault is retried after the interrupt service routine has been executed. For trap conditions, the next instruction in the program being run by the CPU is executed after execution of the interrupt service routine. Abort conditions stop the main program execution entirely necessitating a restart of the program.

Vector	Interrupt
32–255	Available for software and hardware interrupts
17–31	Reserved
16	Coprocessor error
14–15	Reserved
13	General protection fault
12	Stack fault
10–11	Reserved
9	Hardware keyboard
8	Hardware timer
7	Coprocessor not available
6	Invalid opcode
5	Print screen
4	Overflow
3	Breakpoint
2	NMI
1	Debug/single step
0	Divide error

The **IF flag** is used to control whether or not **hardware interrupts** can be processed. When a hardware interrupt is recognised, the CPU clears the IF flag automatically, but this can be reset by the interrupt service routine if additional higher priority hardware interrupts are to be serviced during processing of the interrupt service routine.

The **non-maskable interrupt** (NMI) is a special hardware interrupt that is connected to the NMI pin of the CPU. The NMI is assigned an interrupt number of 2, although, since it cannot be masked by other interrupts, it effectively has the highest priority and is designed to be recognised in the shortest possible time. Conditions such as a power failure or memory read or write errors typically trigger this interrupt.

2.2.13 Memory map

Memory is addressed by reference to segments and offsets. However, as we have seen, the actual value of the segment may be on any 16-bit boundary. Memory itself is divided into **blocks**. There are 16 blocks each of size 64 k (65 536 bytes). Block 0 is the first block and starts at address 00000 and extends to 0FFFF. Block 1 starts at 10000 and goes to 1FFFF etc.

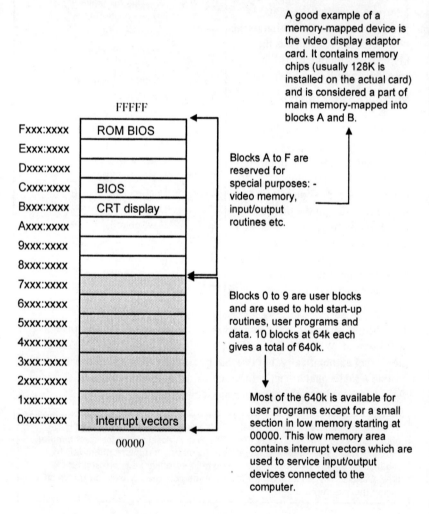

A good example of a memory-mapped device is the video display adaptor card. It contains memory chips (usually 128K is installed on the actual card) and is considered a part of main memory-mapped into blocks A and B.

Blocks A to F are reserved for special purposes: - video memory, input/output routines etc.

Blocks 0 to 9 are user blocks and are used to hold start-up routines, user programs and data. 10 blocks at 64k each gives a total of 640k.

Most of the 640k is available for user programs except for a small section in low memory starting at 00000. This low memory area contains interrupt vectors which are used to service input/output devices connected to the computer.

16 × 64k blocks = 1 MB RAM

2.2.14 Real and protected mode CPU operation

80286 processors and above can operate in either **real** or **protected** mode. When operating in real mode, the CPU can execute the base instruction set of the 8086/8088 processors. In protected mode, the CPU makes use of advanced features for memory management and multi-tasking under the Windows operating system. In protected mode, 80386+ processors can act in virtual 8086 mode allowing 8086 instructions to run in a "DOS" window.

Multitasking under Windows requires programs running on the computer to be isolated from each other and a protected mode of operation is thus required. In protected mode, a program cannot write directly to memory. Instead, any data to be written or read is done to **virtual memory space** and transferred to physical memory using a process called **virtual-to-physical translation** by the CPU.

> Bit 0 in the control register is the protection enable (PE) bit which, in real mode, toggles the CPU into protected mode. At reset, PE is initially 0 and the CPU in real mode. When Windows starts, it toggles this bit to 1 to place the processor into protected mode. The activities in this book are designed for real mode operation of the CPU. This can be simulated in a DOS command window or by starting the computer in "command" mode.

When operating in protected mode, the CPU register structure is different to that used in real mode. The most important additions are **descriptor tables** which hold information about memory and interrupts for each task being run. Details of each task or application being run on the microprocessor are held in the **task register**.

Each task is assigned global and local memory resources. All tasks can access the global address space, but a task cannot access another task's local address space.

Each task is also assigned a **privilege level**. The kernel is responsible for low level tasks such as memory management, I/O and task sequencing. The kernel has the highest privilege level: 0.

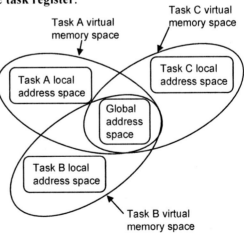

Tasks with a lower privilege level can use routines that have a higher privilege level but cannot modify them. User applications programs are assigned the lowest privilege level: 3.

The combination of local address space and hierarchy of privilege levels allows the instructions and data for all the running tasks to be isolated from each other. Data in one task is thus protected from errors arising in another task.

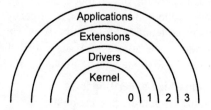

Unlike real mode operation, I/O operations from a user application do not have the required privilege level and so must perform these functions in conjunction with an I/O **device driver** which does. This ensures that I/O is done without violating not only the address space of another running application, but also that the I/O does not adversely affect the low level task sequencing and memory management responsibilities of the CPU.

The significance of protected mode operation for interfacing is that when multiple tasks or applications are being run by the CPU, it appears to the user that they are operating simultaneously whereas task activity is actually **time-shared** within the CPU. For time-critical interfacing applications, the user must be aware of the limitations imposed by this time sharing and hierarchy of privileges. **Virtual device drivers** (VxD) typically have a kernel level of privilege that permits direct I/O and this, together with **direct memory access**, are required for time-critical interface applications.

When a DOS program is run in a DOS or command window within a Windows environment, the CPU is placed into **virtual 8086 mode** by setting the **VM flag** in the extended flag register. The DOS program is still run as a protected mode task, and when the CPU switches to this task, the VM flag is set as part of the task switching process. The DOS mode program is assigned a privilege level of 3. The memory addressing scheme of the task simulates that of a real mode CPU operation and can be configured as part of the task properties.

2.2.15 Review questions

1. Give brief answers to the following questions:

   (a) How many memory locations (or memory cells) can be addressed by a 8086 microprocessor and why?

   (b) What is the largest (unsigned) hexadecimal number that can be stored in one memory location and why?

   (c) How many memory locations can be read in one read/write cycle and why?

   (d) What is a long word and how is it stored in memory?

2. List the four types of special internal registers that exist in the 8086 microprocessor.

3. Explain the difference between a port and an address.

4. Draw a diagram which outlines the main components of a microcomputer system (e.g. the CPU, memory etc). Describe the function of each main component and how each communicates with the others. Indicate also what governs the amount of memory addressable by a program.

5. What is the ROM BIOS?

6. Which bit in the flags register indicates whether or not a subtraction operation produced a negative result.

2.2.16 Activities

1. Start your computer into DOS mode, or open a DOS command window from your Windows environment.

2. Enter the mem command from the DOS prompt and determine how much RAM memory your computer is fitted with and how it has been allocated. Fill in the table below with the values shown on your screen.

3. Start the debug program by typing the command debug at the DOS prompt. The debug prompt is a '-' character and indicates debug is ready for a command.

4. Enter the r command and display the contents of the registers. Fill in the table with the values indicated on your screen (include the last line of the debug output in this table).

```
Memory Type                  Total    Used    Free
----------------             -------- ------- -------
Conventional
Upper
Reserved
Extended (XMS)
----------------             -------- ------- -------
Total memory

Total under 1 MB

Largest executable program size
Largest free upper memory block
```

```
AX=      BX=      CX=      DX=      SP=      BP=      SI=      DI=
DS=      ES=      SS=      CS=      IP=
```

5. What is the status of all the flags? Are they set or reset? Is there any one flag that is set differently to the others? Why would this be?

6. The last line in the register listing displays machine code and assembly language of the instruction pointed to by the CS:IP registers. Compare the segmented address at the beginning of this line with the indicated contents of the CS and IP registers. CS: _____ IP: _____

7. To list the contents of a single register, we enter r **XX** where **XX** is the register name. Debug responds with the current contents of the register and then allows us to change those contents. Display the contents of the AX register and change it to 00FE.

8. To display the flags, we enter r f at the debug prompt. We can then set any of the flags by entering in the appropriate flag code. Display the flags and then change the parity flag to even.

AX	accumulator
BX	base
CX	count
DX	data
CS	code segment
DS	data segment
SS	stack segment
ES	extra segment
IP	instruction pointer
SP	stack pointer
BP	base pointer
SI	source index
DI	destination index

		Flag codes	
Flag		Set	Reset
CF	Carry	CY	NC
PF	Parity	PE	PO
AF	Auxiliary	AC	NA
ZF	Zero	ZR	NZ
SF	Sign	NG	PL
IF	Interrupt	EI	DI
DF	Direction	DN	UP
OF	Overflow	OV	NV

9. The "memory dump" command is d. The syntax is:

    d [address]

   If the segment base address is not specified, then the address is taken to be the offset to the current contents of the DS register. Examine the contents of the BIOS area of memory which is located at F000:0000. Note that the d command displays 128 bytes starting at the address you specify. The d command lists the memory contents in hex and attempts to interpret any ASCII characters and if any are found to be valid ASCII, the characters are displayed on the right-hand side of the screen. Continue to display the contents of the BIOS area of memory until you find the copyright message from the manufacturer of your computer. Record the memory location at which the copyright message appears. _____ : _____

You can continue to press d without any parameters to display more memory contents. Some very interesting messages can be found in the BIOS contents.

10. Debug permits the display, enter, fill, move, compare and search for data in memory. The "enter" command e, allows us to change the contents of memory. The syntax is: e [address] [data].
    Enter the value FF into memory location DS:0000 by typing: e
    DS:0000 FF. Verify that the contents of DS:0000 have changed by entering e ds:0000 and then pressing <enter> key to terminate enter mode.

11. The "fill" command fills a block of memory with all the same values. The syntax is: f [start address] [end address] [data].
    Initialise memory locations DS:0000 to 0100 with zeros. Note: the end address parameter is specified with an offset only and is assumed to be the same as that as the starting address. Verify the contents of these memory locations with the d command.

12. The "move" command allows us to copy a block of memory from one place to another in memory. The syntax is:
    m [start address] [end address] [destination address].
    Again, the segment base address is either implied (the DS register contents) or specified within the start address. Move the contents of DS:0000 to DS:0100 to DS:0200.

13. The "search" command allows us to scan a block of memory and search for a specific byte The syntax is:
    s [start address] [end address] [data].
    The address for any matches is displayed. Search memory locations F000:0000 to F000:FFFF for the characters "read failure".

    > Character data may be entered in a debug command if it is delimited by quote marks. When debug processes the command, the ASCII value of the characters is substituted.

    Display the contents of nearby memory locations with the d command.

14. In addition to loading and running machine language programs (which we will investigate in the next laboratory session), debug also is a handy hexadecimal calculator. The h command allows us to add and subtract hex numbers. Both operations are performed by the same command.
    The syntax is: h [hex number 1] [hex number 2] and the sum and difference of the two numbers is displayed (in hex).
    (a) Calculate the sum of the hex numbers 00FF and AB10.
    (b) Determine the negative of the number A3 (this will be displayed in 2's complement notation).

```
MOV  BL,07H

MOV  DX,02FCH
MOV  AL,0AH
OUT  DX,AL

     DX,02F8H
     AL,0AH
```

2.3 Assembly language

2.3.1 Instruction set

A microprocessor can only act upon instructions which are specified in its **instruction set**. The instruction set consists of a series of hexadecimal codes, or **opcodes**, which are recognisable to the instruction decoder within the CPU. Each series of microprocessor has a unique instruction set, although many instructions are so common that they are found with only minor modifications in all microprocessors. In the 8086 CPU, instructions are 1 to 6 bytes long.

Some common **classes of operations** for which instructions are usually provided are:

- Data movement
- Integer and floating-point arithmetic
- Logical operations
- Shift and rotation of bits
- Bit manipulation
- Program control (branching)

A sequence of opcodes arranged to perform a particular task is called a **machine language** program or just **machine code**. To execute a machine language program, the machine code needs to be stored into the code segment of memory. The first byte of the program is stored at the lowest address and subsequent bytes stored at higher memory addresses in sequence.

Machine code instructions can be from 1 to 6 bytes in length. As an example, the following 3 bytes of code move the literal number 2000H into the AX register:

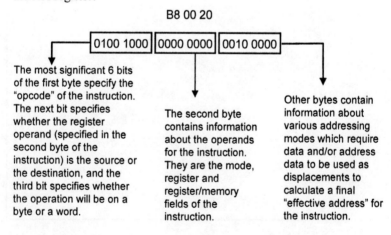

B8 00 20

The most significant 6 bits of the first byte specify the "opcode" of the instruction. The next bit specifies whether the register operand (specified in the second byte of the instruction) is the source or the destination, and the third bit specifies whether the operation will be on a byte or a word.

The second byte contains information about the operands for the instruction. They are the mode, register and register/memory fields of the instruction.

Other bytes contain information about various addressing modes which require data and/or address data to be used as displacements to calculate a final "effective address" for the instruction.

2.3.2 Assembly language

The instruction decoder within the CPU can only interpret machine code instructions which are defined in the instruction set. Machine code programming is extremely laborious and for this reason programs are usually written using **assembly language**. An assembly language program is converted into machine code by an **assembler** program.

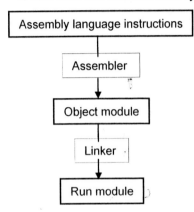

| Assembly language instructions |

| Assembler |

| Object module |

| Linker · |

| Run module |

Source code in ASCII text. Symbolic labels are used in jumps so that absolute addresses need not be calculated by the programmer.

Machine code with all addresses specified relative to a base address. The object module is thus **relocatable**. For large programs, several such object modules may be created.

Separate object modules are combined into a single run module in which cross-references between modules are resolved.

The run module created by the linker is relocatable. The base address to which all other addresses are referenced is supplied by the operating system when it is loaded and run. The linker creates an executable file (with an .exe extension). In some circumstances, a smaller .com file may be made using a relocating loader. The loader provides the base address and then resolves all relative addresses into absolute addresses. The resulting program file is also executable but is given a .com extension. Com files are smaller and execute faster than exe files since the overhead in resolving addresses has been eliminated.

Memory location	Machine language
1F6D:0100	B8
1F6D:0101	00
1F6D:0102	20
1F6D:0103	8E
1F6D:0104	D8
1F6D:0105	BF
1F6D:0106	00
1F6D:0107	00
1F6D:0108	B9
1F6D:0109	FF
1F6D:010A	00
1F6D:010B	BA
1F6D:010C	00
1F6D:010D	00
1F6D:010E	89
1F6D:010F	15
1F6D:0110	47
1F6D:0111	49
1F6D:0112	75
1F6D:0103	F7
1F6D:0114	90

2.3.3 Program execution

The microprocessor can only interpret machine code instructions specified in its instruction set. While running a machine language program, the **program counter** or **instruction pointer** register holds the offset of the address of the next instruction to be executed. The segment base address is held in the **code segment** (CS) register. The initial value of CS and IP is determined by the operating system when the program is run by the user. The following sequence then occurs.

FETCH–EXECUTE cycle

1. CPU **fetches** the instruction at the address given by the program counter (PC register) (and address +1 for instructions which are 2 bytes long) and the instruction pointer register is incremented before the instruction is executed.

The fetch operation consists of the address being put onto the address bus; a read cycle is requested on the control bus. The hex data at locations address (and address+1 for 2 byte instructions) is brought over to the CPU on the data bus to the instruction decoder.

2. The instruction is decoded by the **instruction decoder** which decides what action to take next.

Some instructions require further data to be read from memory (the operands) while others are stand-alone instructions may be acted upon immediately. The instruction decoder determines the set of control signals required for that instruction.

3. Operand data is fetched from memory as required.

4. The instruction is **executed**. Typically, the arithmetic logic unit (ALU) performs the necessary anding and oring etc. The result from the instruction goes into a register or back to RAM memory (which would involve a write cycle). The flags in the **status register** are also set.

The address of the next instruction to be executed is in the IP register (from step 1) and the program cycle starts again. If the program contains some **branching** instructions, then the address to which to branch to is placed in the IP during the execute step ready for the next fetch cycle.

2.3.4 Assembly language program structure

An assembly language program contains statements which may be either **assembly language instructions** (from the instruction set) or **assembler directives** (which are instructions for the assembler program to follow when the program is assembled into machine language).

Assembly language programs use the code segment of memory to hold instructions, the data and extra segment for data, and the stack segment for stack data. To facilitate programming, an assembly language program is also divided into segments.

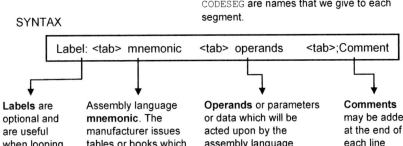

The assembler directive SEGMENT and ENDS define the beginning and end of a program segment. For example, the data segment of a program would look like:

```
DATASEG SEGMENT
... Data definition directives
DATASEG ENDS
```

Following the data segment, the code segment which contains the actual CPU instructions.

```
CODESEG SEGMENT
    ASSUME CS:CODESEG
    PUSH BP
    PUSH DS
    MOV BP,SP
    MOV DS,ES:[SI]
    .
    More instructions....
    .
    POP DS
    POP BP
    RET 08H
CODESEG ENDS
```

Note that each segment of a program begins with a directive SEGMENT and ends with a directive ENDS.

These program statements are written in a text editor and saved as a source file. An assembly language program finishes with the END directive which tells the assembler to stop assembling when it reaches this line in the file.

In these examples, DATASEG and CODESEG are names that we give to each segment.

SYNTAX

Label: <tab> mnemonic	<tab> operands	<tab>;Comment

Labels are optional and are useful when looping back to repeat a section of a program.

Assembly language **mnemonic**. The manufacturer issues tables or books which list all available mnemonics (the instruction set) and their function.

Operands or parameters or data which will be acted upon by the assembly language instruction. For 8088 instructions requiring two operands, the first is the destination and the second is the source.

Comments may be added at the end of each line (preceded by a tab and ";").

2.3.5 Assembler directives

Name <tab> directive   <tab> operands   <tab>;Comment

There are various groups of assembler directives:

1. Symbol definition directives allow names to be assigned to constants,
   addresses, operands etc. There are two definition directives:

   > *name* EQU *expression*
   > *name* = *numeric expression*

2. Data definition directives define memory space for variables.

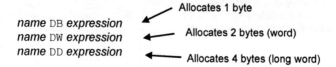

   > *name* DB *expression*   Allocates 1 byte
   > *name* DW *expression*   Allocates 2 bytes (word)
   > *name* DD *expression*   Allocates 4 bytes (long word)

3. External reference directives

   > PUBLIC *expression*   Allows variables and routines which
   > EXTRN *name:type*   exist in other programs to be used in
   > INCLUDE *file name*   the current program.

4. Segment and procedure directives divide the program into segments
   and/or subroutines. The directives SEGMENT and ENDS mark the
   beginning and end of a program segment.

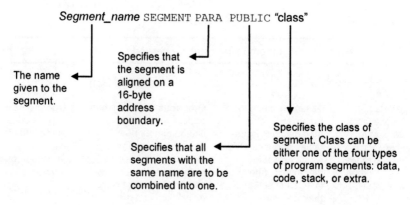

   Segment_name SEGMENT PARA PUBLIC "class"

   The name given to the segment.

   Specifies that the segment is aligned on a 16-byte address boundary.

   Specifies that all segments with the same name are to be combined into one.

   Specifies the class of segment. Class can be either one of the four types of program segments: data, code, stack, or extra.

2.3.6 Code segment

The program segments can be written in the source file in any order. The code segment is perhaps the most important since it contains the actual assembly language statements that are to be executed. An ASSUME directive is usually included in the code segment and is used to assign the segment registers to the base addresses of the program segments.

```
ASSUME DS:name, CS:name, ES:name, SS:name
```

Example:

```
CODESEG SEGMENT 'CODE'
  ASSUME CS:CODESEG,DS:DATASEG
  CODESEG ENDS
END
```

In this example, "CODESEG" is the name given to the code segment. The ASSUME statement specifies that the CS register holds the segment base address for the "CODESEG" segment. The "CODE" class in the SEGMENT statement also identifies the "CODESEG" segment as a code segment.

PROC and ENDP define a portion of code which is used as a subroutine. The last instruction (before ENDP) must be RET. A procedure may be NEAR or FAR. NEAR procedures are defined and called within a single code segment (the CPU needs only to push the return address IP onto the stack while the subroutine executes). For a FAR procedure, both CS and IP are pushed onto the stack.

```
PROCNAME PROC FAR
  . .
  . .
RET 08H
PROCNAME ENDP
```

For procedures that are called from other modules, its name must be declared "public" using the PUBLIC assembler directive.

```
CODESEG SEGMENT "CODE"
ASSUME CS:CODESEG,DS:DATASEG
PUBLIC PROCNAME
```

2.3.7 Assembly language shell program

Here is a useful summary of the statements required to produce a simple
assembly language program which when assembled and linked, can be run
from the DOS prompt.

```
CODE         SEGMENT "CODE"
             ASSUME CS:CODE
MYPROG       PROC FAR
             PUSH DS
             PUSH AX
             .
             .
             . More statements
             .
             RET
MYPROG       ENDP
CODE         ENDS
             END
```

This program shell is a
very simple application of
assembly language
which is suitable for a
very short programs not
requiring the passing of
command line
parameters. It is suitable
for the assembly
language interface to the
serial data acquisition
system to be described
later in this book.

The name of the program is "MYPROG". The statements required to
produce a workable assembly language program depend on the operating
system and the method by which the program is to be run. If the assembly
language program is to be called as a subroutine inside a higher level
program, then the handling of parameters and restoration of the stack is
different to the case where the program is to be compiled into a stand-alone
executable (EXE) file.

The example shown here is suitable for a stand-alone program. The
program is assembled into an **OBJ** file which is then linked to form an
EXE file which can then be executed from the DOS prompt.

Particular care has to be taken with the RET instruction. RET without any
parameters appears sometimes to POP 4 bytes off the stack which is why in
this example, we have pushed DS and AX onto the stack at the start of the
program. If you do not include these statements (PUSH DS, PUSH AX)
your program will hang up and not return to the DOS prompt when
finished.

The END statement is also important if there is more than one procedure in
the program file. If there is more than one PROC and ENDP bracket, then
END must be followed by a label which indicates which procedure is to run
when then program is started.

2.3.8 Branching

Unless otherwise instructed, the CPU will advance from one instruction to the next in sequence. This linear sequence of execution can be varied by **branching**. There are two types of branching.

Unconditional   jump straight to the instruction located at the specified address. The syntax is: JMP *label*

Branching involves an adjustment to the contents of the instruction pointer (IP). To branch to a label, the CPU obtains a number whose value depends on where *label* is located in the program relative to the current IP address. This number (which may be positive or negative) is added to the contents of the IP to give the offset of *label*. This offset is then placed into IP and execution proceeds from CS:IP. If the jump or branch is to a place within the same segment, then it is a NEAR jump. A branch to a different segment is a FAR jump.

Conditional   jump to an instruction at an address, the value of which depends upon the result of a test. Jcc *label*

Bit positions in the flag register indicate the results of the instructions as they are executed. Many instructions cause the flags to be set. Some instructions do not set any flags at all. Program **control statements** test the flags and allow branching to other parts of a program outside the main sequence of instructions.

The symbol cc is a **condition code**. If cc is true, then the program branches to the instruction prefixed *label*. If false, then program execution continues with the next instruction following Jcc.

Jcc	flags tested	
jo	overflow flag is set	OF=1
jno	overflow flag clear	OF=0
jz	equals zero	ZF=1
jnz	not equals zero ZF=0	
jnc	carry flag clear	CF=0
jc	carry flag set	CF=1
js	sign flag set	SF=1
jns	no sign	SF=0

The bit positions, or flags, in the **flag register** are tested and program execution is varied according to the branch instruction.

The condition codes are set by the preceding instruction to the branch. Most assembly language statements set the condition codes as part of the execution procedure within the CPU. Precisely which codes are affected by the execution of a statement depends on the command. The Jcc instruction does not set the flags in the condition code register, it only tests those flags. Thus, one may use several Jcc instructions in sequence, each testing the result of a single previously executed instruction.

2.3.9 Register and immediate addressing

Many assembly language instructions require data to be read or written to memory locations and/or registers. The term **addressing** is used to describe the method by which operands for the source and destination instructions are specified.

Register and **immediate addressing** means that the operand is either a register or specified as a constant within the assembly language instruction itself.

In **register addressing**, an operand is fetched from, or written to, a register. For example:

    MOV AX,DX

In this example, the 16-bit contents of DX are copied into AX. The contents of DX are not changed.

In **immediate addressing**, the actual number specified in the program statement is used as the source operand. For example:

    MOV CX,7FH

In this example, the hex number 7FH is moved into register CX. However, CX is 16 bits wide, and 7F is an 8-bit number: 0111 1111. Since this number is positive (it has a zero as the msb), the most significant 8 bits of CX are filled with zeros.

CX: 0000 0000 0111 1111

For a negative number, e.g. A3, the msb of CX is filled with 1s.

CX: 1111 1111 1010 0011

The MOV instruction

This is one of the most common instructions in an assembly language program. The general syntax of the instruction is:

MOV destination, source

Data is actually copied, not moved, from the source to the destination. The data in the source is not changed. There are some restrictions on MOV:

- Data cannot be moved from one memory location to another by a single MOV command.
- An immediate value cannot be moved into a segment register.
- Data from one segment register cannot be moved into another segment register.
- The CS register cannot be specified as the destination for a MOV instruction.

Some of these restrictions can be avoided by transferring data into a data register (e.g. AX) and then to the desired destination.

2.3.10 Memory addressing

If an operand is stored in memory, then the CPU must calculate the actual physical address from which to read or write the data. The physical address is formed from a segment base address and an offset. The offset is referred to as an **effective address**. The segment base address can be the contents of any of the segment registers. The effective address can be formulated in a variety of ways. In general, the effective address is formed from:

EA = base + index + displacement

The actual physical address is thus:

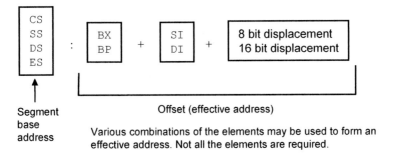

Segment base address

Offset (effective address)

Various combinations of the elements may be used to form an effective address. Not all the elements are required.

In **direct memory addressing**, information about the address is given in the instruction directly. An example is:

```
MOV AX, [0A40H]
```

This example says to move the contents of memory location with offset 0A40 into the AX register. There are several points about this example that require attention. First, there is no segment base address specified in the operand. If this is the case, then the contents of DS are assumed to be the desired segment base address. Second, AX is 16 bits wide, and so a word is moved from the offset and offset+1 with the msb of the AX register receiving the data at offset+1.

The general format for direct memory addressing is: DS:[direct address]

Size codes

There are no explicit size codes used in 8086 assembly language instructions. The size code is taken from the size of the operands. For example, in the MOV instruction, moving data into or out from a segment register is always a word (2-byte) operation. Moving data into or out from AL would be a byte operation since AL is 8 bits wide.

2.3.11 Indirect memory addressing

As with direct addressing, in **indirect addressing**, the effective address
(i.e. offset) is combined with the contents of DS to form an actual
physical address. The effective address is determined from the contents
of either a base or index register.

```
MOV CX, [BX]
```

In this example, the brackets indicate indirect addressing and BX contains
a 16 bit number which is used as a **relative offset** with DS to obtain the
absolute address (and address+1) which contains the 16-bit data to be
moved into CX. Note that the term "relative offset" here means that the
offset is considered to be relative to the contents of the DS register.

The general format is:
$$DS \; : \; [\begin{matrix} BX \\ BP \\ SI \\ DI \end{matrix}]$$

Based addressing is particularly useful for accessing data in tables or
lists. It involves a displacement which is added to the contents of the BX
or BP register to form an effective address.

```
MOV [BX] + 0A10H,AL
```
In this example, the offset for the destination
operand is found by adding the number 0A10H to
the contents of BX. This is then used with the
contents of DS as the base segment address to
form the physical address from which to obtain the
operand data.

The general format is:

$$\boxed{\begin{matrix} DS \\ SS \end{matrix}} \; : [\begin{matrix} BX \\ BP \end{matrix}] \; + \; \boxed{\begin{matrix} \text{8-bit displacement} \\ \text{16-bit displacement} \end{matrix}}$$

If BP is used, then the default register for the segment base address is SS
rather than DS.

Segment base address

In these examples, the default segment base address for offsets (or effective
addresses) is the value in the DS register. However, this can be overridden by
specifying a segment register explicitly. For example:

```
MOV AX,ES:[0A40H]
```

2.3.12 Indexed memory addressing

There are several forms of indexed addressing, all of which use a **displacement** as a pointer to the start of an array of data in memory, and an index register as an index to select a specified element in that array.

```
MOV AL,[SI]+1010H
```

The example here shows a **direct indexing** mode. The displacement 1010H is added to the contents of the stack index register to form an effective address. The default segment base address is given by the contents of the DS register. The advantage of this type of addressing is that the stack index can be incremented or decremented to find the next or previous element in an array of data that begins at DS:1010H.

$$DS : [\begin{array}{c} SI \\ DI \end{array}] \quad + \quad \boxed{\begin{array}{l} \text{8-bit displacement} \\ \text{16-bit displacement} \end{array}}$$

A combination of based addressing and direct indexed addressing results in a **based index** addressing mode. This is useful for accessing two-dimensional (m × n) arrays. The displacement locates the array in memory. The base register specified the m coordinate, and the index register specifies the "n" coordinate of the element.

```
MOV AL,[BX] [SI]+1010H
```

The effective address is found from the contents of BX added to the contents of SI and then added to the value of the direct displacement 1010H. The default segment base address is DS.

$$DS : [\begin{array}{c} BX \\ BP \end{array}][\begin{array}{c} BX \\ BP \end{array}] \quad + \quad \boxed{\begin{array}{l} \text{8-bit displacement} \\ \text{16-bit displacement} \end{array}}$$

2.3.14 Interrupts

Software interrupts are used routinely in assembly language programs to call upon BIOS services to perform basic I/O tasks. Within the CPU processing operation, software interrupts are initiated using the INT statement.

    INT interrupt number

When an interrupt instruction is processed, the following sequence occurs:

1. The flags register is pushed onto the stack
2. Interrupts are disabled to prevent the interrupt routine being interrupted by a lower priority interrupt.
3. Contents of the CS and IP registers are pushed onto the stack
4. The address pointer for the **interrupt service routine** is retrieved from the **interrupt vector table** and loaded into CS and IP registers.
5. The CPU begins executing instructions located at address CS:IP

The following example shows how to read the system clock to obtain the current date and time.

```
MOV AH,2CH      ; SPECIFY DOS SERVICE NUMBER 2C
INT 21H         ; CALL INTERRUPT TO EXECUTE SERVICE
```

Software interrupts are numbered 32 and beyond and are generally assigned a higher priority than external hardware interrupts. Most software interrupts are assigned by the operating system BIOS.

BIOS interrupts

Software interrupts generally offer a service. The service is called by a service code placed in the AH register. Some common software interrupts are:

Interrupt	Function
05h	Print screen
10h	Video service
11h	Equipment list service
12h	Memory size service
13h	Disk drive service
14h	Serial communications service
15h	System services support
16h	Keyboard support service
17h	Parallel printer support services
18h	ROM BASIC
19h	DOS bootstrap routine
1Ah	Real time clock service routines

2.3.15 Review questions

1. What is the relationship between machine language op-codes, mnemonics and the assembler. Also state why you cannot have an assembler that will produce an executable program which will run on more than one type of computer.

2. What is the sequence of events inside the CPU during the execution of a machine language program statement?

3. Determine the physical address given by the following segment:offset 4000H:2H.

4. What is the general syntax of an 8086 assembly language statement?

5. Write a short assembly language program that will arrange two 8-bit numbers in ascending order.

6. The AX register contains the value 1100H and BX contains 2B01H. Write down the contents of the AX register after each of the following assembly language statements executes:

```
AND    AX,BX
OR     AX,BX
XOR    AX,BX
```

7. The following program fragment places a character on the screen. If the hex number A6 is placed in AL, explain what appears on the screen. How would you have an assembly language program display the actual hex number in AL on the screen?

```
MOV    AH,9H
INT    10H
```

2.3.16 Activities

The **debug** program can be used to create a machine language program
from our assembly language input. The command is the assemble or a
command.

1. Start the debug program and enter the **a** command together with a
    starting address as shown: a CS:0100

    Debug responds with the starting address for our program as CS:0100
    and an input prompt _. Enter the short assembly language program
    shown here and press the enter key at the last _ prompt to terminate the
    assemble command.

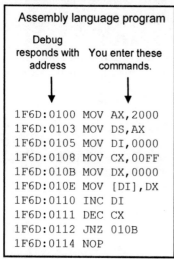

Assembly language program

Debug
responds with   You enter these
address   commands.

```
1F6D:0100 MOV AX,2000
1F6D:0103 MOV DS,AX
1F6D:0105 MOV DI,0000
1F6D:0108 MOV CX,00FF
1F6D:010B MOV DX,0000
1F6D:010E MOV [DI],DX
1F6D:0110 INC DI
1F6D:0111 DEC CX
1F6D:0112 JNZ 010B
1F6D:0114 NOP
```

This program fills memory locations
2000:0000 to 2000:00FF with zeros.

2. Verify the contents of the
    assembly language program
    using the unassemble **u**
    command: u CS:0100

3. The program is now ready to be
    executed. The go command
    executes the entire program, while
    the trace command executes a
    specified number of lines and then
    stop, whereupon the contents of
    registers and memory locations can
    be examined before proceeding.
    Enter the trace t command:

    t =CS:0100

    Examine the contents of the
    registers after each statement. Press
    t to continue execution until the
    last instruction is processed – do
    not press t after the NOP step.

Questions:
(a) Examine the contents of memory locations 2000:0000 to 2000:00FF and
    check their contents.
(b) Examine the contents of the AX, DX, DI, CX registers and explain their
    values.
(c) Unassemble the program and determine the machine language code for
    the first move statement.
(d) In the program, the JNZ statement was followed by a hex number 010B.
    What is the significance of this number?

4. The trace command is useful for executing the program one line at a time. The go command executes the program without pausing. However, we must be careful only to execute the instructions that we have entered into the memory locations CS:0100 to CS:0114. The go command allows us to execute a block of statements by specifying the memory location at which to stop processing. Change the program slightly by editing memory location 010B to:

```
MOV DX,00FF
```

By using the **a** command to reassemble this line:

```
a CS:010B
```

Run the program using the go command:

```
g =CS:0100 0114
```

Examine the contents of 2000:0000 to 2000:00FF and comment.

5. It is customary to use a text editor to write large assembly language programs. The program is saved to a disk file and then translated into machine code using an **assembler**. Using a text editor, create a text file containing the assembly language statements as shown.

Note: "Zero" is a name we assign to the procedure. We only need this name if we link this program with a series of others. Other procedures can then call this program by its procedure name.

```
CODESEG SEGMENT 'CODE"
        ASSUME CS:CODESEG
ZERO    PROC FAR
        PUSH DS
        MOV AX,0H
        PUSH AX
        MOV AX,2000H
        MOV DS,AX
        MOV DI,0H
        MOV CX,00FFH
START:  MOV DX,0H
        MOV [DI],DX
        INC DI
        DEC CX
        JNZ START
        RET
ZERO    ENDP
CODESEG ENDS
        END
```

6. Save the file to disk with an .asm file name extension. For example: LAB2.ASM

7. Start the Microsoft Macro Assembler by typing in masm at the DOS prompt. The assembler responds with prompts for the source file, object file, source listing file, and cross-reference listing file. The default names for the object file and the cross-reference file can be selected by just pressing the enter key at the prompt. For this exercise, do not accept the default NUL.LST (which produces no list file) but enter the name: LAB2.LST.

8. If there are any syntax errors in the source file, they will be reported by the assembler. If there are no errors, then proceed. Edit the source listing file created by the assembler and note its contents. The source listing shows both the source and corresponding machine code instructions.

9. The object file created by the assembler contains machine code but this code is not yet in executable form. A separate program called a "linker" is used to create the final executable program file. Start the linker program by typing in link at the DOS prompt. The linker asks for the object file names (of which there is just the one in our present exercise) and the name of the final run file, a linker map file, and any library files. Enter the name of the object file (e.g. LAB2.OBJ) and also specify a map file with the same file name but with a MAP extension. No library files are required so simply press enter at the LIB and DEF prompts. (Ignore the linker warning about there being no stack segment.)

10. The linker creates two files, an executable program file with an EXE extension, and a map file which contains the start address, stop address and program length for each program segment used by the program. Examine the contents of the MAP file using a text editor and note the stop address of the program.

11. The EXE file used by the linker is in executable form and can be executed directly from the DOS prompt. However, we shall run the program from within the DEBUG environment in this exercise. Assuming that the executable file is called LAB2.EXE, start the debug program with this file name as a command line parameter:

```
debug lab2.exe
```

Note the status of the registers (use the r command).

12. Verification that the program has been loaded correctly can be done by using the unassemble command. The starting address is given by CS:0000. Unassemble commands from this starting address to the stop address given by the MAP file (see Step 10).

    `u CS:0000 stop address`

    Compare with the source listing (LAB2.ASM).

13. Now run the program using the go command. Use the memory dump command d to verify that the program has performed its intended purpose.

    `g =CS:0000`

    What message does debug display after executing the program?

Questions:

(a) Examine the LST file created by the assembler and compare the machine language output with that produced by the unassemble command of debug (see Step 2).

(b) When running this assembled and linked program from within debug using the go command, we did not need to specify a stop address as we did in Step 4. Why?

(c) In the source file (.ASM) we used a label as the target for the JNZ command. Examine the unassembled program (from Step 12) and verify that the assembler calculated the correct offset for this jump.

---

14. Write an assembly language program that will swap the contents of locations 2000:0000 and 2000:00FF.

15. Write an assembly language program to perform an 8-bit subtraction. The contents of 2000:0000 are to be taken away from the contents of 2000:0001 and the answer stored in 2000:0002. Compare your answer to a pen and paper check using 2's complement notation. Make sure you include an example that gives a negative result.

16. Write an assembly language program which will add the two 16-bit
    numbers below and store the result in locations 2000:0000, 2000:0001
    and 2000:0002 (you'll need a third memory location since the addition
    of the two numbers shown below will overflow 16-bit positions).

    1 0 1 1 1 0 1 1 1 0 1 0 1 0 1 0
    1 1 0 1 1 1 0 1 1 1 0 0 1 1 0 0

17. Write an assembly language program which will inspect the contents of
    the AX register and either increment, decrement, or leave unchanged
    the data located at address 2000:0000 depending on whether the data in
    AX is positive, negative or zero (in 2's complement notation).

18. Write an assembly language program which will fill the locations
    2000:0000 to 2000:FFFF with zeros. (You can use a branch statement
    to loop back until some condition is met – for example, load the number
    0000 into a data register and increment that number each time you fill a
    memory location until the number equals FFFF. Try using the J[cc]
    commands to set this up).

19. Write an assembly language program that will swap the contents of
    location 2000:0400 to 2000:04FF with the contents of 2000:0500 to
    2000:05FF.

20. Write an assembly language program to scan the memory locations
    2000:0000 to 2000:FFFF to see if one of them contains a byte equal to
    a byte stored in the AL register. Put the address of that location in the
    BX address register.

21. Write an assembly language program that will return the contents of a
    memory location whose segment base address is specified in DS and
    offset in SI. The contents are to be loaded into the lower half of AX.

22. Write an assembly language program that will return call DOS
    service 21C via interrupt 21H. Examine the contents of CH, CL, DH
    and DL and determine what information is being returned by this
    service. (The DOS service number is to be placed into AH before
    interrupt is called.)

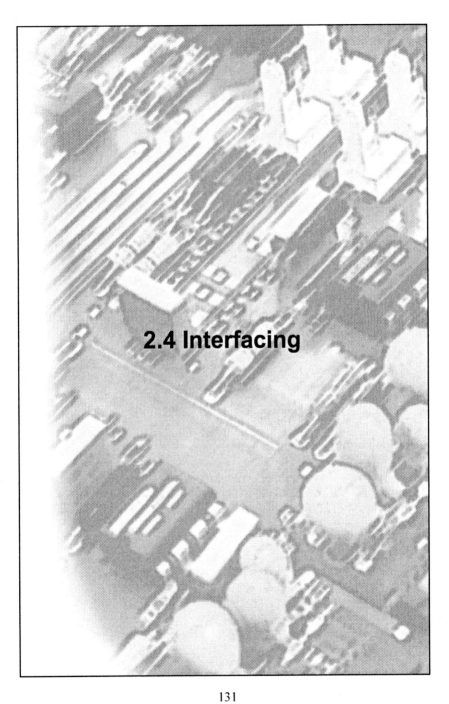

2.4 Interfacing

2.4.1 Interfacing

The term **interfacing** is used to describe the connection between a transducer or some other external device and the microcomputer. Interfacing circuits may be required to deal with various levels of incompatibility:

- incompatible voltage levels
- changing current levels
- electrical isolation
- timing of data transfers
- digital to analog and analog to digital conversions

Mechanical or electronic devices requiring connection to the microprocessor unit can be anything from the output screen or monitor, the keyboard, and external instruments. Two main problems are usually encountered when interfacing such devices:

- Most devices do not operate at the same **speed** as the microprocessor.
- There may be more than one device which requires **servicing** at any one time.

A **port** is a connection from the outside world to the microprocessor. The purpose of an **input port** is to transfer information from the outside world to the microprocessor. An **output port** provides information to the outside world from the microprocessor. Each port has an address and is thus connected to the address bus and information to or from the port is transmitted over the data bus. From the microprocessor's point of view, a port is very similar to a location in memory.

Typical I/O ports on a microcomputer

Ports may be **memory mapped**, interfacing direct to the computer's RAM, or be assigned a separate **port address**. An interface adaptor connects the data bus to the I/O device using compatible signals when the port is accessed by the CPU for read/write operations.

2.4.2 Input/Output ports

In an 8086-based microcomputer, I/O ports are identified using a 16-bit **port number** or **port address**. Thus, there are a total of 65 536 possible ports numbered 0 to 65 535 (FFFF). The CPU uses a signal on the control bus to specify that the information on the address bus and data bus refers to a port and not a regular memory location. The port with the specified number then receives or transmits the data from its own internal memory.

Input ports generally require servicing (i.e. their data to be read) at irregular intervals and further, their signals may only appear momentarily. Techniques such as **polling**, **interrupts** and **direct memory access** are used to service ports as required.

Port number	I/O device
0000 – 001F	Direct memory access controller
0020 – 003F	Programmable interrupt controller
0040 – 005F	System timer
0060 – 0060	Standard 101/102-key keyboard
0061 – 0061	System speaker
0062 – 0063	System board extension for ACPI BIOS
0064 – 0064	Standard 101/102-key keyboard
0065 – 006F	System board extension for ACPI BIOS
0070 – 007F	System CMOS/real time clock
0080 – 009F	Direct memory access controller
00A0 – 00BF	Programmable interrupt controller
00C0 – 00DF	Direct memory access controller
00E0 – 00EF	System board extension for ACPI BIOS
00F0 – 00FF	Numeric data processor
0170 – 0177	Intel(R) 82801BA Ultra ATA storage controller – 244B
0170 – 0177	Secondary IDE controller (dual fifo)
01F0 – 01F7	Intel(R) 82801BA Ultra ATA storage controller – 244B
01F0 – 01F7	Primary IDE controller (dual fifo)
02F8 – 02FF	Communications port (COM2)
0376 – 0376	Intel(R) 82801BA Ultra ATA storage controller – 244B
0376 – 0376	Secondary IDE controller (dual fifo)
0378 – 037F	ECP printer port (LPT1)
03B0 – 03BB	Intel(r) 82815 graphics controller
03C0 – 03DF	Intel(r) 82815 graphics controller
03F0 – 03F5	Standard floppy disk controller
03F6 – 03F6	Intel(R) 82801BA Ultra ATA storage controller – 244B
03F6 – 03F6	Primary IDE controller (dual fifo)
03F7 – 03F7	Standard floppy disk controller
03F8 – 03FF	Communications port (COM1)
04D0 – 04D1	Programmable interrupt controller

2.4.3 Polling

The easiest method of determining when a device requires servicing is to ask it. This is called **polling**. In this method, the CPU continually and sequentially interrogates each device. If a device requires servicing, then the request (or bus access) is granted. If the device does not require servicing, then CPU interrogates the next device.

Polling is very CPU intensive since the processor must spend a large amount of time interrogating devices which do not require servicing. However, the procedure may be easily implemented in software making it flexible and convenient. In some circumstances, polling may be actually faster than more direct methods of interfacing (interrupts and DMA).

Interfacing in a multitasking environment

Interfacing in a multitasking operating system like Windows brings with it many issues that may require special attention. The three main methods of obtaining data from an interfaced device (polling, interrupts, and DMA) cannot be guaranteed to occur at a particular time. This causes problems for time-critical applications in which the time at which the data is recorded is important, and also for applications requiring large amounts of data to be rapidly collected.

Steps can be taken to minimise the problems. I/O devices such as general purpose data acquisition cards make use of virtual device drivers employing commands with low level privileges, hardware buffering, and bus-mastering DMA can be used with some effect but cannot remove the limitations of the overall system placed on it by the multitasking environment.

In some applications, where the cost and effort is appropriate, interfacing can be done at the transducer and the data buffered and transferred to the microcomputer at a time convenient to the microprocessor. In these systems, the transducer contains a microprocessor of its own and is programmed using an erasable programmable read only memory (EPROM). Commands can be sent to the on-board microprocessor to run different internal programs using an ordinary serial communications protocol.

Intelligent transducers contain all the power to obtain the necessary data from the sensor under a variety of conditions, report error conditions and self-calibrate under the control of a supervisory computer via an internet, radio or direct cable connection.

2.4.4 Interrupts

Hardware interrupts are controlled by
the 8259 **programmable interrupt
controller.** I/O devices managed by
hardware interrupts are printers, keyboard,
and disk drives. The **IRQ** allocation is a
hardware device interrupt number
simply used to conveniently label the
devices making use of the 8259 controller.
The lower the IRQ, the higher the priority.

For interfacing applications, the time taken to register and process an
interrupt (**interrupt latency**) can lead to the need for the I/O device to
be heavily buffered. In addition, time critical interfacing applications
may not work as desired.

Typical IRQ allocations

0	System timer
1	Standard 101/102-keyboard
2	**Programmable interrupt controller**
3	Communications port (COM2)
4	Communications port (COM1)
5	(free)
6	Standard floppy disk controller
7	ECP printer port (LPT1)

Additional interrupts from
2nd 8259 controller

8	System CMOS/real time clock
9	Intel(r) 82815 graphics controller
9	ACPI IRQ holder for PCI IRQ steering
9	SCI IRQ used by ACPI bus
10	SoundMAX integrated digital audio
10	Intel(R) 82801BA/BAM SM bus controller – 2443
10	ACPI IRQ holder for PCI IRQ steering
11	Intel(R) 82801BA/BAM USB universal host controller – 2444
11	3Com 3C920 integrated fast ethernet controller
11	ACPI IRQ holder for PCI IRQ steering
12	PS/2 compatible mouse port
13	Numeric data processor
14	Primary IDE controller (dual fifo)
14	Intel(R) 82801BA Ultra ATA storage controller – 244B
15	Intel(R) 82801BA Ultra ATA storage controller – 244B

PCI-based systems are able to share IRQ assignments. When a **shared interrupt** is
activated, the operating system calls each of the assigned interrupt service routines
until one of the routines (configured by the device driver of the hardware) claims the
interrupt by conducting its own tests. For example, often registers are available
within each device that can identify whether the device has signalled an interrupt

2.4.5 Direct memory access (DMA)

In normal data transfer, data is transferred from one memory location to another through **registers** in the CPU. The CPU has to hold the data temporarily while it switches the control bus signal from a read to a write since the data bus cannot be in a read state and a write state at the same time. This temporary storage of data and resulting transfers into and out of the CPU is time consuming and wasteful for interfacing applications that require rapid accumulation of data and precise timing.

In **direct memory access** or **DMA**, data can be transferred directly between memory and the I/O port since I/O memory locations are independent of RAM memory. DMA requires full control of the address, data and control buses. When a DMA transfer is to occur, a **DMA controller** 8237 IC requests control of the bus from the CPU. The CPU promptly grants control and suspends any bus-related activity of its own. The DMA controller then transfers data from port to memory, or memory to port directly, without any stack or register overhead operations that would normally be required by the CPU to accomplish the same task. The DMA acts as a **third party** to the data transfer. The **latency time** associated with DMA transfer is only a few CPU cycles.

The 8237 DMA controller has a number of independent **channels**, each of which is assigned to a particular device. Channel 2 is usually assigned to the floppy disk controller. DMA can take place as a single byte or word, a **block** of bytes, or on demand up to a set number of bytes. DMA transfers can be initiated by a hardware request (via DREQ input on the 8237) or a software request using a **request register**.

With a PCI bus, DMA management can be performed not only with the DMA controller, but also by the device requiring DMA access. In such systems, the device that gains control of the bus is called the **bus master**. For interfacing applications, the combination of **bus-mastering DMA** and a high speed PCI bus ensures that data transfer occurs as fast as possible from the I/O device to memory. Further, bus-mastering DMA does not require the allocation and usage of DMA channels since the DMA controller is not involved. Bus-mastering DMA is referred to as **first party** DMA since the I/O device itself is handling all the data transfer.

2.4.6 Serial port

Most microcomputers are fitted with one and often two **serial ports**. These serial ports are labelled COM1 and COM2. The numbers 1 and 2 are for our "external" convenience only. The actual "internal" port numbers or addresses are 3F8 for COM1 and 2F8 for COM2.

The COM ports can usually be found on the back panel of a microcomputer and may take the form of either 25 or 9 pin connectors. These pins are connected to **buffers** which convert the pin voltages used for data transmission over external cables (usually using the **RS232** standard) to TTL levels used for data transfer within the computer. The internal signals are generated by a special communications IC called a **UART**.

The serial port is most often used for data communications. Hence, one of the signal lines carries data either being transmitted from, or received by, the computer. The other signals are used to control the flow of data and to establish a communications link between the two serial ports on two different computers. Often, the serial ports are connected by a **modem** which converts digital data into analog signals for transmission over a telephone line.

The handling and control of transmission is done by setting and reading the binary data which appears in the internal registers of the UART. Each of these registers has an address (i.e. the **port address**) in the port address space of the computer.

The 9 pin connector was introduced to save space when the parallel and serial ports were placed on a single interface card.

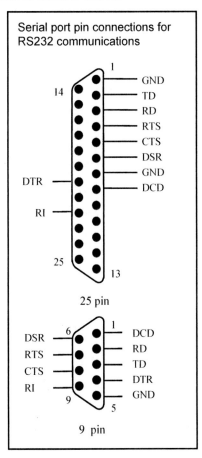

Serial port pin connections for RS232 communications

25 pin

9 pin

2.4.7 Serial port addresses

The port addresses for IBM compatible microcomputers have been
standardised for many years.

Purpose	COM1	COM2	
Tx,Rx data	3F8	2F8	◄──── Base address
Interrupt enable	3F9	2F9	It is customary to refer to the
Interrupt ident	3FA	2FA	first address as the "base
Line control	3FB	2FB	address".
Modem control	3FC	2FC	
Line status	3FD	2FD	
Modem status	3FE	2FE	

In Windows, it is easy to obtain the base
address for the COM ports. In Control Panel,
select System, and then "Device Manager".
Select COM ports and then properties.

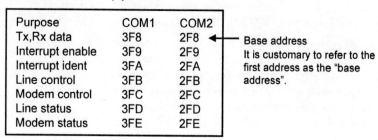

Each port address is a
register that allows the serial
port to be initialised and
operated on by software
commands. That is, the serial
port controller ship, the 8250
UART, is **programmable** in
the sense that its operation
can be controlled by
software rather than hard-
wired circuitry.

When a serial port interface
card is added to a computer,
the **base address** must be
set, either by a jumper on the
card, or by software. This
allows the card to be
configured as COM1 or
COM2 (or even COM3 or
COM4) as desired.

2.4.8 Serial port registers

LCR (Line Control Register)

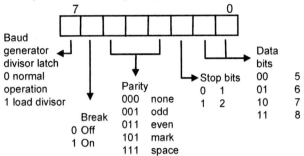

LSR (Line Status Register)

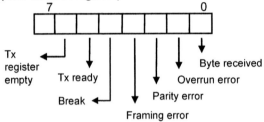

MSR (Modem Status Register)

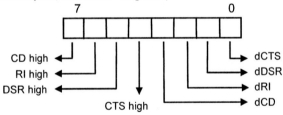

The d flags are set if the state of the control lines has changed since they were last read.

The MCR is not set by the 8250 UART itself. We must set bits in it to control the UART operation and/or the modem control lines.

MCR (Modem Control Register)

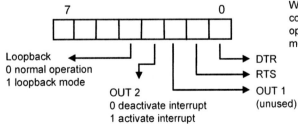

2.4.9 Serial port registers and interrupts

It is most common to operate the serial port (i.e. such as the **8250 UART**) through the use of interrupts. However, this need not always be the case. The 8250 has four internal interrupt signals which can be connected through to the CPU's IRQ interrupt line via an INTR pin on the UART. The OUT2 bit in the Modem Control Register specifies whether or not to connect the UART INTR output to the CPU's IRQ line. In this way, the internal interrupts generated by the UART can be optionally used by the CPU.

Note: COM1 usually uses IRQ4 and COM2 IRQ3 on the CPU.

A 1 in the corresponding bit position enables the internal interrupt. This will not be registered at the CPU IRQ line unless OUT2 in the MCR is also set to 1.

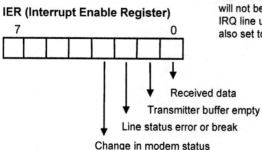

IER (Interrupt Enable Register)

Received data
Transmitter buffer empty
Line status error or break
Change in modem status

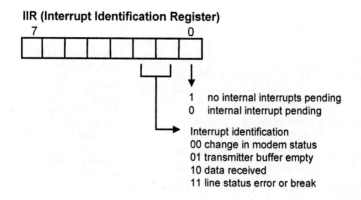

IIR (Interrupt Identification Register)

1   no internal interrupts pending
0   internal interrupt pending

Interrupt identification
00 change in modem status
01 transmitter buffer empty
10 data received
11 line status error or break

2.4.10 Serial port baud rate

The **baud rate** is a measure of the number of bits per second that can be transmitted or received by the UART. This rate is regulated by a clock circuit which, for most UARTS, is on the chip itself and can be programmed.

Thus,

$$B = \frac{1.8432 \times 10^6}{16D}$$

$$D = \frac{115200}{B}$$

where D is called the baud rate divisor and must be loaded into the UART.

Example:
A baud rate of 9600 is required. What is the divisor D?

$$D = \frac{115\,200}{9600}$$

$$= 12$$

How is this divisor loaded?

1. Bit 7 of the LCR must be set to 1.
2. The LSB of D is written to the port base address (e.g. 3F8 for COM1).
3. The MSB of D is written to the port base address +1 (e.g. 3F9).
4. Bit 7 of LCR is cleared (and perhaps also set for other parameters such as baud rate, stop bits etc).
5. Check port base address +1 for the desired interrupt settings.

The **UART clock** must operate at 16 times the desired baud rate. The clock is based around the operation of a crystal oscillator which, in the case of a 8250 UART, is set to a constant 1.8432 MHz. This clock signal is stepped down through a series of counters to obtain the desired clock rate for the chip to give the desired baud rate.

2.4.11 Serial port operation

Although it is possible to write and read from the serial port registers
directly, it is more convenient to use either applications' program
languages or BIOS **service routines**. Most applications' languages have
statements or functions available which facilitate the programming of the
serial port. For example, the

```
OPEN "COM1:9600,N,8,1" AS #1
```

statement in BASIC allows the serial port to be configured without a
detailed knowledge of the actual port addresses. However, for interfacing
applications, direct manipulation of the registers is required. For example,
the BIOS service routines on an IBM compatible PC do not provide a way
to set RTS for hardware handshaking.

In Visual Basic, it is necessary to make use of the **serial port object**.

MSComm1 has properties that can be set in code that allow the
serial port to which is is assigned to be configured.

```
.MSComm1.Settings = "9600,E,7,1"
.MSComm1.InputLen = 0
.MSComm1.RTSEnable = True
.MSComm1.DTREnable = False
.MSComm1.PortOpen = True
```

These high level instructions ultimately result in a series of assembly
language instructions which call BIOS service routines through the
interrupt system.

The serial port initialisation
parameters are: baud rate, parity, stop
bits, data bits. They are combined into
an 8-bit number which is loaded into
AL prior to calling the interrupt.

The service to be called (0 for
initialise serial port) is placed into
AH. Parameters for the service are
placed in AL. The interrupt is
called, and the results placed in AL
(or AX for service 3), e.g.:

```
mov   AH,0
int   14H
```

2.4.12 Parallel printer port

The **parallel port** normally found on microcomputers is generally used for printer output although there are some input lines which are used to report printer status (such as paper out etc.). The **Centronics** printer interface consists of 8 data lines, a data strobe, and acknowledge, three control and four status lines.

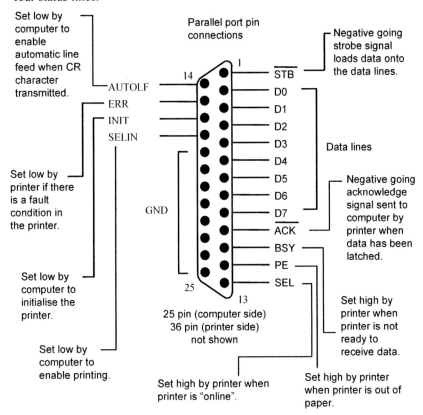

The printer port is driven by the parallel port adaptor. In the adaptor, there are three registers which are assigned I/O port addresses. The byte to be printed is held in the data register which is at the port base address. The printer status register contains the information sent to the computer by the printer, and has an address of base+1. The printer control register has address base+2 and contains the bit settings for computer control of printer functions.

2.4.13 Parallel port registers

Printer port data register (base)   378

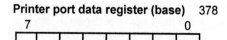

Printer port status register (base+1)   379

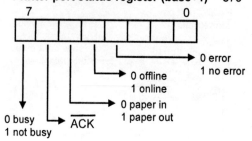

Printer port control register (base +2)   37A

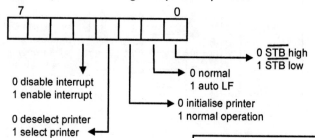

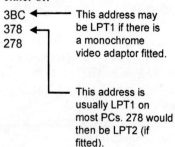

The base address for the parallel printer port can be either of:

3BC ← This address may
378 ← be LPT1 if there is
278   a monochrome
       video adaptor fitted.

This address is usually LPT1 on most PCs. 278 would then be LPT2 (if fitted).

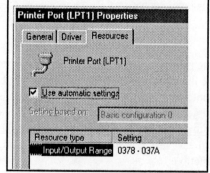

In Windows, the base address of the parallel port is obtained through the Device Manager in the Control Panel. Select LPT1 and then Properties.

2.4.14 Parallel printer port operation

Although it is possible to write directly to the parallel printer port registers, it is customary to use the **BIOS service routines** available through the computer's operating system. Mostly this is done indirectly through high level program statements like PRINT. However, it is possible (and sometimes desirable) to call the BIOS routines directly from an assembly language program.

BIOS routines are called through interrupt 17H. Three services are available and are selected by the value placed in AH. For writing a byte to the printer, the data to be printed is put into AL. The DX register is set to indicate the LPT port to use (0 for LPT1:).

AH	BIOS service
00	Write byte
01	Initialise printer
10	Report printer status

After the service has been executed, the contents of the printer status register are reported in AL.

Centronics type parallel connector

When the **printer port** is being used through the BIOS service routines or being accessed directly, the following sequence is required to write the data:

- The data to be written is placed in the Printer Port Data Register. That is, the byte is written to the printer port base address.

- The readiness of the printer to accept data is confirmed by testing the bits in the Printer Port Status Register.

- The STB line is then pulsed low by writing a 1 to bit 0 of the Printer Port Control Register. This transfers the data from the Printer Port Data Register to the Data Lines on the port connector.

Note: Although the parallel printer port is usually used for printing (i.e. output) there is no rule against using the port for input and output for other peripherals. The more recently introduced **IEEE 1284** (1994) standard defines five modes of data transfer: Compatibility Mode (standard mode); Nibble Mode (4 bits in parallel using status lines for data); Byte Mode (8 bits in parallel using data lines); EPP (Enhanced Parallel Port – used primarily for CD-ROM, tape, hard drive, network adapters, etc.) and ECP Extended Capability Port – used primarily by new generations of printers and scanners.

2.4.15 Review questions

1. Arrange the following statements, which describe the sequence of events when a CPU services an interrupt-driven device, in the correct order.

   (a) The return address (i.e. contents of the program counter) is placed on the stack.
   (b) The CPU is directed to the interrupt service routine.
   (c) The CPU returns to the main program.
   (d) The interrupt service routine is executed.
   (e) The CPU checks the interrupt mask.
   (f) The return address is put back into the program counter PC.

2. Briefly describe the difference between programmed and interrupt-driven I/O.

3. What are the three different types of interrupts in an 8086 CPU-based computer?

4. How many channels are offered by an 8259 DMA controller? How can further channels be accommodated?

5. What is the difference between bus-mastering DMA and DMA via 8259?

6. What should the contents of the Line Control Register be for a UART to be configured for 7 data bits, 1 stop bit, and even parity?

7. What is the main difference between serial and parallel communications? Give examples of advantages and disadvantages of each method.

8. How is data usually communicated out through the parallel port?

9. For interfacing applications, what limits the maximum speed of data acquisition in polling, interrupts, and DMA?

2.5 A to D and D to A conversions

2.5.1 Interfacing

Interfacing to a microcomputer is the process whereby the physical phenomenon to be measured is converted into an analog electrical signal by a transducer. The signal is then digitised by an **analog to digital converter (ADC)** and then stored in memory.

The ADC can be located either near the transducer, or, as is more common, on a special purpose interface card installed inside the microcomputer. The interface card interfaces directly to memory either by DMA or as a memory-mapped device.

When a digital number is to be displayed on some external device, it is converted into an analog electrical signal by an interface adaptor containing a **digital to analog converter (DAC)**. The actuator then converts the analog signal into a physical quantity.

Physical phenomena:
Temperature
Voltage
Position
Velocity
Force...

Transducer
(sensor and signal conditioning)

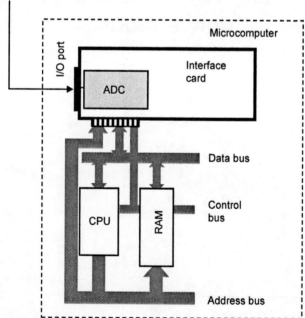

2.5.2 The Nyquist criterion

An analog signal varies
continuously with time. A
good example is a sine
wave of frequency ω.

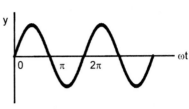

If we were to store this wave as a sequence of numerical data, we would
choose data pairs (y,t) at convenient intervals. The smaller the interval, the
more accurate the representation of the original signal when we come to
reconstruct it from the data.

Reconstruction

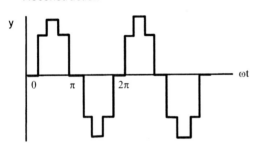

Discrete
numerical data

0	0
π/4	0.797
π/2	1
3π/4	0.707
π	0
5π/4	−0.707
3π/2	−1
7π/4	−0.707
2π	0

In this example, data was sampled at π/4 or 45°
intervals. If we decreased the sampling interval to π/8 (i.e. an
increase in the sampling rate), then we would obtain a more
accurate representation of the original analog curve.

What minimum **sampling rate** is required to reconstruct the signal? The
Nyquist criterion states that the sampling rate (samples per second)
should be greater than twice the highest frequency component (cycles per
second) of the signal.

If the signal was sampled at
intervals greater than π, then
the resultant wave, when
reconstructed, may still be a
sine wave but at a different
(lower) frequency. This is
called **aliasing**. The
reconstructed signal is an alias
for the original signal.

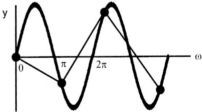

2.5.3 Resolution and quantisation noise

Now, a very interesting problem arises when analog data is to be stored in a digital system like a computer. The problem is that the data can only be represented by the range of numbers allowed for by the analog to digital converter. For example, for an 8-bit ADC, then the magnitude of the full range of the original analog data would have to be distributed between the binary numbers 00000000 to 11111111, or 0 to 255 in decimal. Numbers that don't fit exactly with an 8-bit integer have to be rounded up or down to the nearest one and then stored.

If the ADC were able to accept input voltages from 0 to 5 volts, then full scale, or 5 volts, on the input would correspond to the number 255 on the output. The resolution of the ADC would be:

$$\frac{5}{255} = 0.0196 \text{ V}$$

Thus, for an input range of 0 to 5 volts, in this example, the resolution of the ADC would be 19.6 mV per bit or 0.39%. A 12-bit ADC would have a **resolution** of 1.22 mV per bit (0.02%) since it may divide the 5 volts into 4095 steps rather than 255.

In general, the resolution of an N bit ADC is:

$$\Delta = \frac{V_{ref}}{2^N}$$

where V_{ref} is the range of input for the ADC in volts.

Quantisation noise

The **quantisation error** Δe is ± half a bit (LSB) and describes the inherent fundamental error associated with the process of dividing a continuous analog signal into a finite number of bits.

$$\Delta e = \frac{V_{ref}}{2\left(2^N\right)}$$

The quantisation error is random, in that rounding up or down of the signal will occur with equal probability. This randomness leads to the digital signal containing **quantisation noise**, of a fixed amplitude, and a uniform spread of frequencies. The rms value in volts of the quantisation noise signal is given by:

$$e_{rms} = \frac{\Delta}{\sqrt{12}} \quad \longleftarrow \text{resolution}$$

The quantization noise level places a limit on the signal to noise ratio achievable with a particular ADC.

2.5.4 Oversampling

In general, an input signal is comprised of a range of individual or component frequencies. These signals can be separated by Fourier analysis. The range of component frequencies able to be handled by a particular analog to digital converter circuit is called the **bandwidth**.

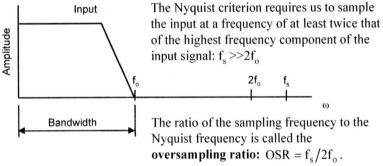

The Nyquist criterion requires us to sample the input at a frequency of at least twice that of the highest frequency component of the input signal: $f_s >> 2f_o$

The ratio of the sampling frequency to the Nyquist frequency is called the **oversampling ratio**: $OSR = f_s/2f_o$.

Even if the Nyquist criterion is satisfied, then the existence of the **quantisation noise** limits the ability of the system to represent the original input signal exactly. This noise, the amplitude of which is independent of the frequency of the signal, becomes a component part of the sampled data. The power associated with this noise P_n is found by integrating e_{rms} over the frequency range 0 to f_o to give:

$$P_n = e_{rms}^2 (1/OSR)$$

where $e_{rms} = \dfrac{V_{ref}}{2^N \sqrt{12}}$

The SNR is:

$$SNR = \frac{P_s}{P_n} \quad \begin{array}{l} \nearrow \text{Signal power} \propto V_{in}^2 \\ \\ \nwarrow \text{Noise power} \end{array}$$

$$SNR_{db} = 20 \log_{10} \left| \frac{V_{ref}}{2\sqrt{2}} \frac{1}{e_{rms}} \right|$$

$$= 6.02N + 1.76$$

The significance of this is that the **signal-to-noise ratio** SNR can be improved by increasing the OSR or increasing N.

Now, the noise voltage is expressed here as an rms value, thus, we must also express the input voltage as an rms value. The maximum SNR is obtained when the full range of the ADC is used. Allowing for both positive and negative halves of the input cycle, the maximum value of V_{in} to the ADC is $V_{in} = V_{ref}/2$, and thus, the rms value of this is: $V_{ref}/2\sqrt{2}$.

Increasing the number of bits increases the signal to noise ratio. However, oversampling with an N-bit ADC also reduces the noise power and thus causes the N-bit ADC perform as if it were an N+w bit ADC. If f_s is the original sampling frequency, then to obtain w extra bits of resolution, the new (or oversampling) frequency is given by $4^w f_s$.

2.5.5 Analog to digital converters

An **analog to digital converter** accepts an input voltage and issues a
positive integer on its output whose binary value is proportional to the
magnitude of the input voltage.

Digital data can be readily stored
and processed on a microcomputer.
Analog signals cannot.

- **Staircase** (or **integrating**) method
- **Successive approximation** method

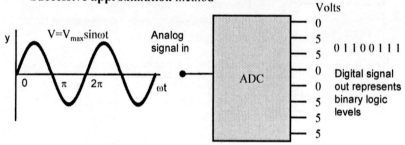

Typical conversion times:

	8 bit	12 bit	16 bit
Integrating	20 msec	40 msec	250 msec
Successive approximation	10 μsec	20 μsec	500 μsec

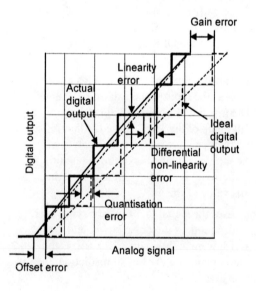

- Linearity error typically ±1/2LSB
- Differential error typically ±1/2 LSB
- Gain error % adjustable by external resistor
- Offset error: adjustable by external resistor

2.5.6 ADC (integrating method)

A to D conversions are usually performed by comparing the unknown input signal voltage to an internal **reference voltage**. A voltage generator supplies a reference voltage which is adjusted until it equals (to within some predefined tolerance level) the input signal voltage.

The reference voltage is linearly increased in small steps until it equals or exceeds the signal voltage and a digital counter is used to record the number of voltage steps tested during the conversion time. The digital count is thus an indication of the magnitude of the voltage input.

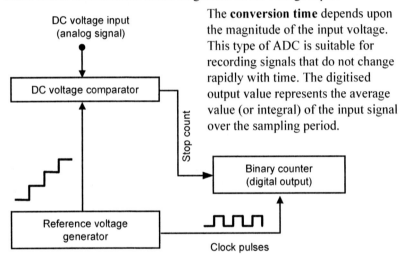

The **conversion time** depends upon the magnitude of the input voltage. This type of ADC is suitable for recording signals that do not change rapidly with time. The digitised output value represents the average value (or integral) of the input signal over the sampling period.

In more sophisticated devices, a **dual slope** technique is used. After an initial zeroing period, the analog input signal is **integrated** (added together) for a fixed number of clock cycles. The integrator input is then connected to an internal reference voltage that has a polarity opposite to the analog input signal.

The number of clock cycles for the integrator to "discharge" to zero is proportional to the magnitude of the original analog signal voltage. The accuracy of the ADC is thus dependent on the accuracy of the reference voltage.

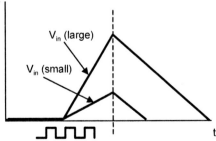

2.5.7 ADC (successive approximation)

In this method, the input voltage is compared to half the full scale voltage and then lower values in succession. The steps are:

1. Set all bits set to zero.
2. Set msb to 1.
3. If V_{in} > D/A, then V_{in} is above half of full scale of output, thus, keep a 1 in msb. If not, then clear msb to zero.
4. Set next msb to 1 (i.e. could have 1100 0000 or 0100 0000).
5. If V_{in} > D/A, then V_{in} is above 7/8 of fsd, thus, keep 1, else clear bit.
6. Set next msb to 1 (e.g. 0010 0000, 0110 0000, 1010 0000 or 1110 0000)
7. If V_{in} > D/A then V_{in} is above 6/8 of fsd, thus, keep 1, else, clear bit position.

and so on until V_{in} is tested against lsb.

The **conversion time** is fixed and is equal to: $\dfrac{N}{f}$

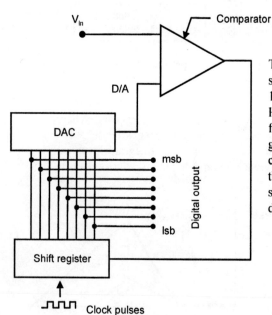

This method allows high speed data acquisition (up to 100 000 conversions/sec). However, the opportunity for errors to be introduced is greater. The fixed conversion time means that the input signal needs to be steady or at least captured during the sampling period.

2.5.8 Aperture error

We have seen that the conversion of an analog signal to a digital output takes time: the **conversion time**, which in the case of a successive approximation ADC, is fixed. Now, if the analog input signal is changing during the conversion time, then the converted output will be in error. This is known as **aperture error**.

For example, for an 8-bit ADC, the smallest increment δ of input signal registered by a single bit will be: $\delta = 1/2^8 = 0.0039$ fraction of full scale of signal.

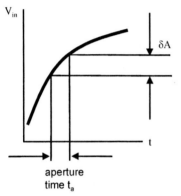

aperture
time t_a

This would be the maximum change in V_{in} for a particular time increment Δt.

During conversion time, the signal changes. For there to be no error in the digitised output, this change must be less than the smallest increment registered by a single bit: i.e. the product $(\delta)(A)$.

Consider a sine wave:

$$V_{in} = A \sin \omega t$$

$$\frac{dV_{in}}{dt} = A\omega \cos \omega t$$

$$\frac{\Delta V_{in}}{dt}\bigg|_{max} = A\omega$$

Maximum rate of change when $\cos \omega t = 1$.

$$\Delta V_{in} = (A\omega)\Delta t$$

$$\Delta V_{in} = \delta A \;@\; \Delta t = t_a$$

$$\delta A = (A\omega)t_a$$

The aperture time is the maximum time interval within which the conversion must occur before the signal changes enough to affect the digitised output.

$$t_a = \frac{\delta}{\omega} = \frac{\delta}{2\pi f}$$

Max aperture time for sine wave of frequency f Hz. Conversion time has to be less than this for no aperture error.

For there to be no error in the output, $\Delta V_{in} < \delta A$. The maximum aperture time is that at $\Delta V_{in} = \delta A$.

e.g. For 100 Hz input, 8-bit ADC, $t_a = 6.2$ μsec.

2.5.9 ADC08xx chip

The ADC08xx series of ICs are 8-bit analog to digital converters which use the successive approximation technique.

The **conversion time** is given by the clock frequency. It takes approximately 64 clock cycles to perform one 8-bit conversion. Thus, to obtain a sampling rate of say 10 000 samples per second, the clock frequency needs to be set to:

$$f = (64)(10 \times 10^3)$$
$$= 640 \text{ kHz}$$

The conversion time is thus:

$$T = \frac{64}{640 \times 10^3}$$
$$= 100 \text{ } \mu\text{s}$$

64 clock cycles for one conversion, 640 × 10³ clock cycles per second.

Conversions are initiated by an external start pulse at pin WR. Conversions begin when WR goes from low to high. WR must remain high during conversion.

When conversion is completed, INTR goes low indicating that the digital data on the outputs is complete and the device is ready for the next conversion.

The chip takes the difference between the two analog inputs as its input signal. If $-V_{in}$ is tied to ground, then the other may be used as a single-ended input. If no external reference voltage is applied at $V_{ref}/2$, then the chip uses an internal reference which depends on the value of V_{cc}.

There is an RC oscillator built into the chip whose frequency is approximately 1/RC. For example:

$$680 \text{ kHz} = \frac{1}{10k(147 \text{ pF})}$$

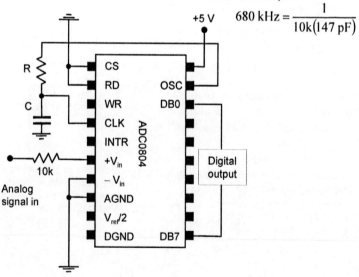

2.5.10 Sample-and-hold

To avoid aperture error, the conversion time and the desired performance characteristics of the ADC circuit must be taken into consideration. For example, given a conversion time of say 100 μsec, what is the maximum frequency of sine wave that can be sampled by the 8-bit ADC0804 without aperture error?

For an 8-bit ADC, $\delta = 1/2^8 = 0.0039$

$$t_a = \frac{\delta}{2\pi f}$$

$$100 \times 10^{-6} = \frac{0.0039}{2\pi f}$$

$$f = 6.2 \text{ Hz}$$

Q. How many bits per second need to be transmitted for this frequency of sine wave? N = $1/620 \times 10^{-6}$ = 1611 bytes/sec = 12 888 bits/sec

Not very high! What to do?

We need a circuit that will take a sample of the input voltage at a particular instant, and hold it until the ADC has processed the conversion - a **sample-and-hold** circuit. There are a number of pre-packaged ICs available, a common one being the LF198, 298, 398 series.

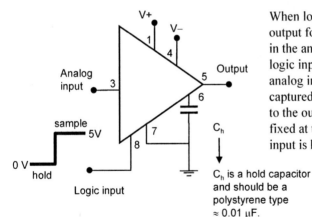

When logic input is high, output follows any changes in the analog input. When logic input goes low, the analog input signal is captured and passed through to the output. Output remains fixed at this value while logic input is held low.

C_h is a hold capacitor and should be a polystyrene type ≈ 0.01 μF.

The time taken for the sample-and-hold circuit to sample the signal and hold it must be shorter than the conversion time (otherwise we wouldn't need to use the circuit!). The above circuit has a conversion time of about 10 μsec.

2.5.11 Sample-and-hold control

Now, to control a sample-and-hold circuit, signals from the ADC
can be used.

WR: Standby mode when WR is
low. Conversions begin when
WR goes from low to high. WR
must remain high during
conversion.

INTR: Usually high. When
conversion is completed, INTR
goes low for eight clocks. WR
must then be held low for about
500 nsec before going high to
initiate a new conversion.

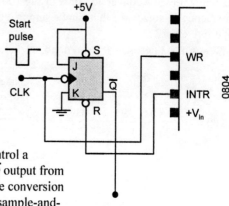

We can use a 7476 **latch** to control a
sample-and-hold circuit. The $\overline{Q}$ output from
the latch can be made low while conversion
is in progress thus sending the sample-and-
hold circuit to "hold".

To sample-and-
hold logic input

CLK1		K1
S1		Q1
R1		$\overline{Q1}$
J1	7476	GND
V_{cc}		K2
CLK2		Q2
S2		$\overline{Q2}$
R2		J2

Action tables:

Normal operation

R	S	Q	$\overline{Q}$
1	1	no change	
1	0	1	0
0	1	0	1

J	K	Clock pulse 1 to 0
0	0	no change, $Q_{n+1} = Q_n$
0	1	$Q_{n+1} = 0$ (RESET)
1	0	$Q_{n+1} = 1$ (SET)
1	1	$Q_{n+1} = \overline{Q_n}$ toggle

- Start pulse initiates conversion
  since it is connected directly to
  WR.
- Since $\overline{SET}$ is high, and initially
  $\overline{RESET}$ is high, the output $\overline{Q}$
  will respond to the clock going
  low and since J is high (with K
  low), $\overline{Q}$ is sent low. Sample
  will be latched on clock signal
  going low. Conversion actually
  begins when clock (WR) goes
  back high.
- At the end of conversion, INTR
  goes low which sends a low to
  $\overline{RESET}$ sending $\overline{Q}$ high
  independent of the signal at J
  and sending sample and hold to
  "sample".

2.5.12 Digital to analog conversion

Digital to analog conversions can be performed using resistor networks and the conversion to an analog signal is usually in the order of nanoseconds. Since the digital information is a step approximation of the input signal, the resulting output from a D to A converter reflects this step nature of the signal.

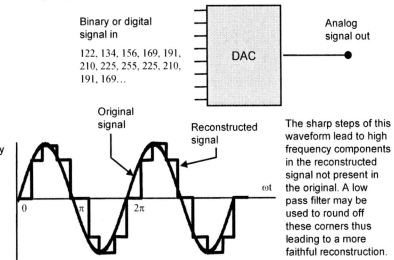

Binary or digital signal in

122, 134, 156, 169, 191, 210, 225, 255, 225, 210, 191, 169...

DAC

Analog signal out

Original signal

Reconstructed signal

y

ωt

0   π 2π

The sharp steps of this waveform lead to high frequency components in the reconstructed signal not present in the original. A low pass filter may be used to round off these corners thus leading to a more faithful reconstruction.

Digital to analog conversions may be made using a **ladder network** of resistors or a **weighted input** to a summing amplifier. The voltage on the output depends upon the voltages applied to the inputs. These voltages may be either 0 (for logic 0) or some supply voltage V_{cc} (for logic 1). The TTL input connected to the lowest value resistor carries more weight than the others, thus, a larger binary or digital input results in a larger analog output voltage.

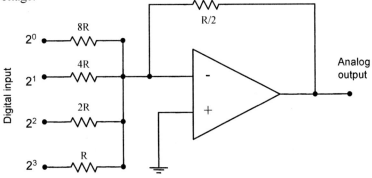

Digital input

2^0 8R

2^1 4R

2^2 2R

2^3 R

R/2

−

+

Analog output

2.5.13 DAC0800

A popular all-purpose 8-bit D to A converter IC is the DAC080x series.
The settling time is in the order of 100 ns.

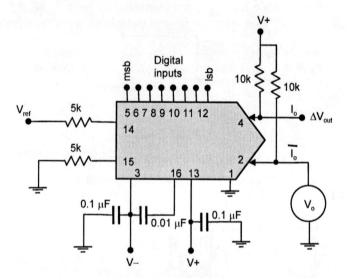

The output for this IC is in the form of two **complementary currents** I_o
and $\overline{I}_o$. In the diagram above, these current outputs are connected to a V+
supply through two 10K resistors. A voltage output can be obtained by
measuring the voltage between the two output terminals or measuring the
voltage of one of the outputs with respect to ground. As the binary value of
the digital inputs increases, I_o increases and $\overline{I}_o$ decreases. A decrease in $\overline{I}_o$
means an decrease in the voltage drop across the 10k resistor and an
increase in V_o measured w.r.t. ground.

V_{ref} provides a current
reference. Setting V_{ref} to
V+ makes V_o swing
positive and negative.
Setting V_o to V+/2 gives
a 0 to V+ analog output.

V_{LC}		CMP
$\overline{I}_{out}$		V_{ref}^-
V-		V_{ref}^+
I_{out}	DAC0800	V+
B1		B8
B2		B7
B3		B6
B4		B5

2.5.14 Data acquisition board

In most circumstances, one would accomplish most interfacing tasks with a general-purpose **data acquisition board**. Such boards fit into an ISA or PCI slot in a microcomputer and would typically contain:

- Sixteen analog input channels (ADC)
- Forty digital I/O
- Four digital output channels (DAC)
- Four 16-bit counter input
- Two 16-bit timer outputs

Multiplexing is used on the input to a single ADC chip to allow multiple and continuous scans of the analog inputs. Interface boards generally allow the 16 analog inputs to be open-ended, or paired to form eight differential inputs. The analog inputs may be fitted with simultaneous **sample-and-hold** circuits to reduce the error associated with sequential sampling of the inputs by the multiplexer.

Configuration and control of interface cards is done using applications program interface (**API**) calls. These are functions provided by the card manufacturer to perform tasks such as data acquisition, counter and timer operation and selection of trigger method.

For multiple channel data acquisition, the **scan rate** (1/scan interval) is an important parameter. Scan rates and other time critical functions are referenced to an on-board clock or an external trigger signal.

Data transfer can be initiated by software polling, interrupts or DMA. **Interrupt latency**, especially under a multitasking operating system, can limit the maximum data transfer rate. Maximum data transfer rate is usually obtained using **bus-mastering DMA**.

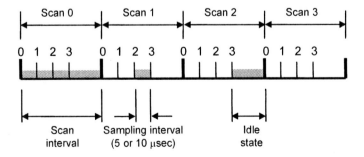

2.5.15 Review questions

1. A 10-bit ADC accepts an input voltage from 0 to 5 V. Determine the resolution of the ADC.   Ans: 4.88 × 10⁻³ V

2. For an input signal of 500 Hz frequency, determine the aperture time of an 8-bit successive approximation ADC.   Ans: 1.24 × 10⁻⁶ s

3. Given a conversion time of say 50 µsec, what is the maximum frequency of sine wave that can be sampled by an 8 bit ADC without aperture error?   Ans: 12.4 Hz

4. What is the purpose of a sample-and-hold circuit?

5. Calculate the settling time of a DAC that is required to convert 16 bit signals from a compact disc player to produce sound in the audible frequency range (say 20–20 000 Hz) whose signals were encoded without aperture error.   Ans: 1.2 × 10⁻¹⁰ s

6. The input signal below is presented to a sample-and-hold circuit. Sketch the output signal which is in turn presented to the ADC when the logic input signal goes from sample to hold as shown.

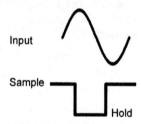

7. Design (in principle) a D to A converter that uses a network of resistors without any active components.

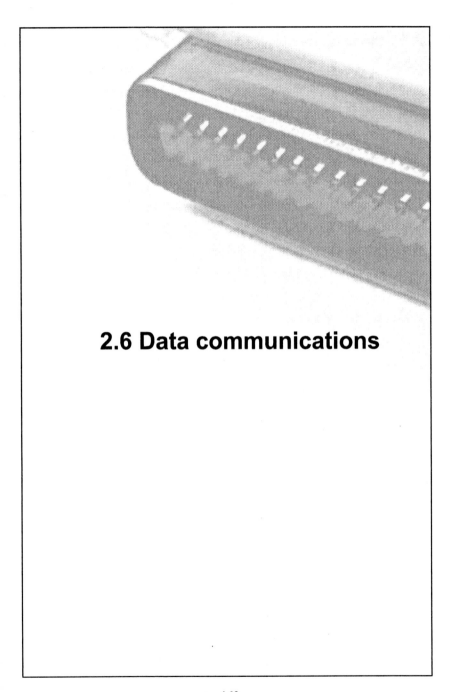

2.6 Data communications

2.6.1 Communications

Once an analog signal has been digitised by the ADC, the digital
information must then be passed to a port of a microcomputer for
subsequent placement on the data bus. An inexpensive, readily available
method is by serial communications using the **serial port**. Other common
methods are to use an ADC **interface card** that interfaces directly to the
computer bus system, and the **GPIB** parallel data bus.

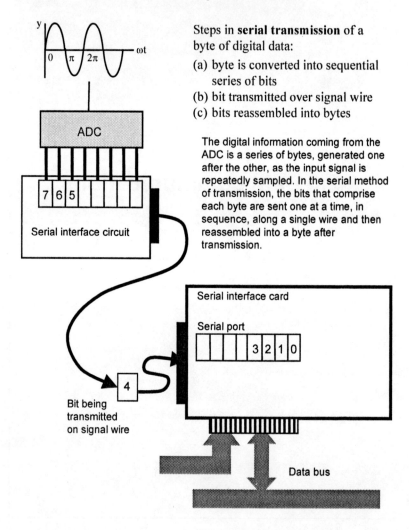

Steps in **serial transmission** of a
byte of digital data:

(a) byte is converted into sequential
series of bits

(b) bit transmitted over signal wire

(c) bits reassembled into bytes

The digital information coming from the
ADC is a series of bytes, generated one
after the other, as the input signal is
repeatedly sampled. In the serial method
of transmission, the bits that comprise
each byte are sent one at a time, in
sequence, along a single wire and then
reassembled into a byte after
transmission.

2.6.2 Byte to serial conversion

At the transmission end of the process, the byte of data is sent bitwise over a signal wire. This is done using a **shift register** in a special integrated circuit called a **UART** (Universal Asynchronous Receiver/Transmitter).

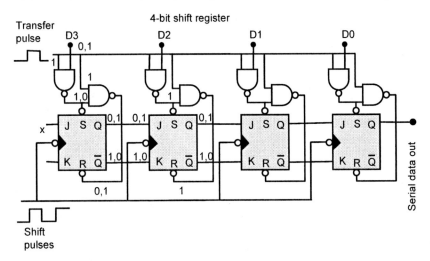

A shift register can be made using cascaded JK flip-flops. A positive **transfer** pulse loads the **asynchronous** inputs R and S with the data to be shifted. For example, if D = 0, then on the transfer pulse, S = 1, R = 0 and Q = 0. If D = 1, then S = 0, R = 1 and Q = 1. Thus, after the transfer pulse, the parallel data is transferred to the J input of each flip-flop. When the transfer pulse goes low, R and S are both at 1 which is the no change state, and on the clock or shift pulse, data is transferred along the chain from Q to J, the serial output bit stream appearing as the last output Q.

At the receiving end, the serial bit stream is converted to parallel 8-bit data by a reverse of the above. Data is clocked in, and then a transfer pulse transfers the data to the parallel outputs.

Now, for long distances, the bit stream is not usually transmitted directly over wires since the binary signals are easily distorted with distance. This, together with the requirement that the transmitter and receiver need to be synchronised, means that alternative arrangements are required to actually transmit the data over distances of more than about 15 m.

2.6.3 RS232 interface

The **RS232** interface standard defines the necessary control signals and data lines to enable information to be transmitted between computer equipment (or data terminal equipment **DTE**) and the modem (or data communication equipment **DCE**). The **modulated carrier signal** is transmitted over a two-wire telephone network by the connecting **modems**.

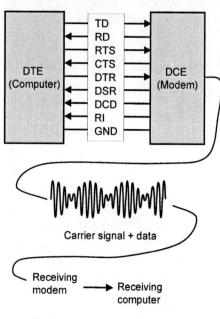

Carrier signal + data

Receiving modem → Receiving computer

1. RTS is raised high by the sending computer indicating that data is ready to be sent.
2. The transmitting modem sends a carrier signal to the receiving modem which raises DCD on its connection to the receiving computer which is thus notified of the existence of a carrier signal.
3. The transmitting modem waits for a preset period for the receiving computer to get ready to receive data.
4. The transmitting modem raises CTS signalling to the sending computer that it may now begin to send data.
5. The transmitting modem receives digital data from the sending computer on the TD line and modulates the carrier wave accordingly. The receiving modem demodulates the carrier and puts digital data on the RD line connection to the receiving computer.
6. When the sending computer has finished sending the data, it clears the RTS signal to the transmitting modem which then drops the carrier and clears its CTS signal. The receiving modem detects loss of carrier signal and it drops its DCD line to the receiving computer.

A modem converts or "**modulates**" digital information into a form suitable for transmission over the telephone network. The receiving modem demodulates the signal back into digital data for use by the receiving computer. Modems transmit information using a sine wave **carrier** which is modulated (either through amplitude, frequency or phase) to carry binary information.

2.6.4 Synchronisation

Start and **stop bits** frame the data bits. When the signal line is not sending data, it is idle and held at **mark** or logic high. The start bit is low or **space** and thus when the receiver sees a transition from mark to space, it knows that the next bit to be received is the lsb of the data being transmitted. After all the data bits have been received, the receiver interprets the next bit to be a stop bit which is mark. If the bit actually received is not mark, then the receiver knows that an error has occurred.

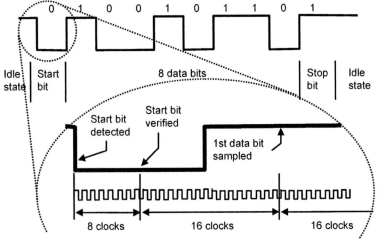

The receiver clock is usually made 16 times the bit rate. When a mark to space transition is detected, the receiver counts 8 clocks. If the signal is still at space, then it is assumed that the signal is a valid start bit. The receiver counts off another 16 clocks and then samples the data until all the data bits and stop bit(s) have been received. The bit sampling thus takes place in the centre of the signal levels.

After the data bit and before the stop bit, there may be a **parity** bit which is used as a check for data validity. In even parity, the total number of 1s (including the parity bit) is made to be an even number (and vice versa). The receiver computes its own parity bit when a byte is received and compares it with that appended by the sender. A mismatch indicates a **parity error**.

The rate of transmission is the data bit rate which is called the **baud** rate. Generally, the baud rate is the same as the **bit rate**; however, some transmission systems are capable of sending more than one data bit (e.g. amplitude and phase modulation) into a transmission bit and the bit rate is thus higher than the baud rate.

2.6.5 UART (6402)

The 6402 UART takes in parallel byte data into the Transmitter Buffer
Register at the inputs TBR1–8. This data is transferred into the Transmitter
Register for shifting. The output bit stream appears at Transmitter Register
Output (TRO). Similarly, serial data is read in at Receive Register Input,
converted to a character in the Receive Register and appears at Receive
Buffer Register outputs RBR1–8. The transmitter part of the circuit
automatically adds start, stop and parity bits according to logic levels
applied at control inputs. The receiver checks for parity and stop bit errors
and issues logic levels at various indicator outputs.

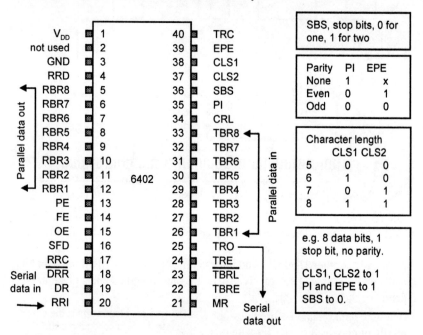

When TBRL goes low, data is read from the input pins TBR1–8 and
transferred to the Transmitter Buffer Register. When TBRL goes from low to
high, data is transferred from the Transmitter Buffer Register to the
Transmitter Register whereupon data is shifted and transmitted out on TRO.

If RRD is held low, then the data on RBR1–8 is that last read from the input
stream at RRI. A high level on DR indicates that the data read is available at
RBR1–8. Once read, DR needs to be reset by a negative pulse to DRR.

A description of the pin functions on the 6402 UART is given below:

RRD: Receiver Register Disable. A high level forces RBR1 to RBR8 to a high impedance state.

RBR8–RBR1: Receiver Buffer Register outputs. Parallel byte data is output here.

PE: Parity Error: High level indicates that the received parity does not match the parity set by the control bits. When the parity is inhibited (PI) this pin is held low.

FE: Framing Error: High level indicates that the first stop bit is invalid.

OE: Overrun Error: High level indicates data received flag was not cleared before the last character was transferred to the Receiver Buffer Register.

SFD: Status Flag Disable: High level input forces the outputs PE, FE, OE, DR, TBRE to high impedance.

$\overline{RRC}$: Receiver Register Clock. Set to 16 times the baud rate.

$\overline{DRR}$: Data Received Reset. A low level clears the output DR to low.

DR: Data Received. A high level indicates character has been received and transferred to the Receiver Buffer Registers.

RRI: Receiver Register Input. Serial data is clocked into the Receiver Buffer Registers from here.

MR: Master Reset. High level clears PE, FE, OE and DR and sets TRE, TBRE and TRO to high. MR should be pulsed high after power-up to reset the UART.

TBRE: Transmitter Buffer Register Empty indicates that the Transmitter Buffer Register has transferred its data to the Transmitter Register and is ready to accept new data.

$\overline{TBRL}$: Transmitter Buffer Register Load. A low level transfers data from the inputs TBR1–8 to the Transmitter Buffer Register.

TRE: Transmitter Register Empty. A high level indicates that transmission of a character, including stop bits, has been completed and that the Transmitter Register is now empty.

TRO: Transmitter Register Output. Serial data output line.

TBR1–8: Transmitter Buffer Register inputs. Parallel data is loaded into the Transmitter Buffer Registers at these inputs.

CRL: Control Register Load. High level loads the Control Register with parity, character length and other settings.

TRC: Transmitter Register Clock. Set to 16 times the baud rate.

The more versatile 8250 (16450) UART extends the functionality of the basic 6402 by having programmable registers that set the baud rate, parity and stop bits, and an interrupt controller. High level commands in an application program set the appropriate bits in the internal registers (see Section 2.4.5). Later 16550 UARTS feature first-in, first-out (FIFO) buffers which allow data transfer to happen at maximum speed while the processor is momentarily occupied by other tasks.

2.6.7 Line drivers

RRI and TRO on the 6402 UART are the serial input and output lines. The **polarity** of the signals is +5 V for **mark** or high, and 0 V for **space**, or low. However, the transmission of serial data along the wire in an RS232 transmission interface requires −15 to −12 V for mark, and +12 to +15 V for space. **Line drivers** are used to convert the logic levels required by the UART to those required at the RS232 interface pins TD and RD.

A very useful IC is the 232CPE dual RS232 **transmitter/receiver**. This IC requires a single +5 V supply and generates +10 V and −10 V necessary to drive the RS232 signal lines. (Note: Handshaking signals CTS, RTS, DSR etc are also at +10 V, −10 V.)

The 232 takes either TTL or CMOS logic levels as inputs, and provides ±10 V at the RS232 outputs. It also receives RS232 inputs, and provides equivalent TTL or CMOS levels as an output. There are two separate channels available for both receiving and transmitting.

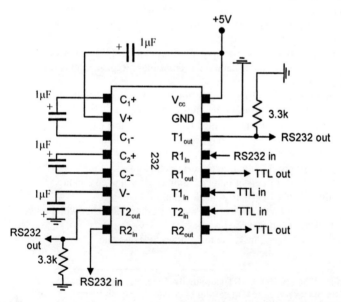

Transmitting: TTL input, RS232 output
Receiving: RS232 input, TTL output
The external capacitors are used by the
internal voltage doublers to obtain ±10 V
RS232 signals.

2.6.8 UART clock

The 6402 has two separate clock inputs but normally, these are driven by the same clock so that the transmit and receive baud rates are the same. The clock frequency is to be 16 times the baud rate. The 6402 chip does not have a programmable **baud rate divider** (as in the 8520) so the desired clock frequency must be supplied externally using a binary counter.

A **crystal oscillator** can be used in conjunction with a high speed inverter to produce a square wave output which can be stepped down to the desired frequency with a binary counter.

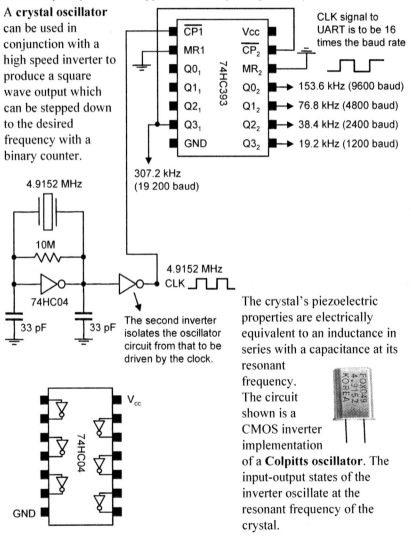

CLK signal to UART is to be 16 times the baud rate

153.6 kHz (9600 baud)
76.8 kHz (4800 baud)
38.4 kHz (2400 baud)
19.2 kHz (1200 baud)

307.2 kHz (19 200 baud)

4.9152 MHz

10M

74HC04

33 pF   33 pF

4.9152 MHz
CLK

The second inverter isolates the oscillator circuit from that to be driven by the clock.

74HCO4

V_{cc}

GND

The crystal's piezoelectric properties are electrically equivalent to an inductance in series with a capacitance at its resonant frequency. The circuit shown is a CMOS inverter implementation of a **Colpitts oscillator**. The input-output states of the inverter oscillate at the resonant frequency of the crystal.

2.6.9 UART Master Reset

MR: Master Reset. High level clears PE, FE, OE and DR and sets TRE, TBRE and TRO to high. MR should be pulsed high after power-up to reset the UART. A simple delay circuit using a capacitor and a NAND gate can be used to send a short positive pulse to MR on power-up. The time constant is simply the product RC, and values of R_1 = 100 kΩ and C = 4.7 μF give a time constant of about 0.5 sec.

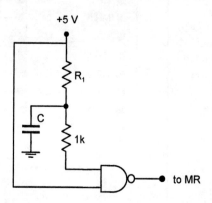

Initially, NAND output is at 0 V. When V_{cc} is applied, the NAND output goes to 1 since one input is at +5 V and the other at 0 V. As the capacitor charges up through R_1, voltage at upper input to the NAND rises and after a time characterised by R_1C, the NAND gate flips from 1 V to 0 V. The 1 kΩ resistor limits the current into the NAND for protection.

A suitable IC is the 7400 series NAND.

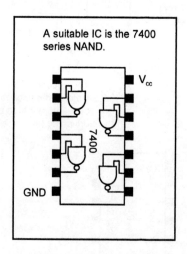

2.6.10 Null modem

The RS232 serial protocol was designed to transmit data over a considerable distance using the telephone network but may also be used for local communication to and from a device attached to the serial port of a computer. The exact same control lines may be used to regulate the flow of data between the connected equipment without the use of a modem; however, there is a problem: if both devices are connected pin to pin and attempt to send data over the transmit line, then no signals will appear on the receive lines. Thus, the transmit and receive lines must be crossed over. This type of connection is called a **null modem**.

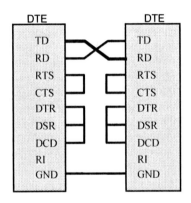

Tying DTR, DSR and DCD together in effect tells the computer that the modem is connected to a telephone line upon which there is a valid carrier signal and data can be either sent or received. Of course, there is no modem present at all, hence the term null modem. TD, RD and signal ground are the minimum requirements for a serial connection between two computers, the other connections may be required if the communications software on the computers tests the modem status before sending or transmitting data.

In the example above, the connections between RTS and CTS, and DTR, DSR and DCD, cause the computer to regard the connection as occurring between modems even if no modem is used at all. For instance, the computer which is sending data first raises its RTS line, which is now directly connected to CTS. The sending computer thus immediately receives a CTS signal from its own RTS and begins to transmit data on TD. The receiving computer receives this data on its RD line.

2.6.11 Serial port BIOS services

BIOS services may be used to initialise and use the serial port. These services are available through interrupt 20 (14H). Parameters for the interrupt are specified in the AL register. The serial port is specified in DX. The four services available are:

0   initialise serial port ──────▶ The serial port initialisation parameters are: baud rate, parity, stop bits, data bits. They are combined into an 8-bit number which is loaded into AL.

1   send one character

The 8-bit data to be transmitted is placed in AL. After transmission, a status code is placed in AH.

Bits	baud	Bits	Parity	Bit	No.
7 6 5	rate	4 3		2	stop bits
0 0 0	110	0 0	none	0	one
0 0 1	150	0 1	Odd	1	two
0 1 0	300	1 0	None		
0 1 1	600	1 1	Odd	Bit	data
1 0 0	1200			1 0	bits
1 0 1	2400			0 0	unused
1 1 0	4800			0 1	unused
1 1 1	9600			1 0	7
				1 1	8

2   receive one character

The received character is placed in AL. A code is placed in AH to report status.

3   read serial port status

The status returned by services 0–2 and that reported by service 3 are in the form of a bit pattern in the AH and AX registers respectively. A 1 in any bit position indicates the condition or error returned.

AH		AL	
7	Time out	7	Receive signal detect
6	Transfer shift register empty	6	Ring indicator
5	Transfer holding register empty	5	Data set ready
4	Break-detect error	4	Clear to send
3	Framing error	3	Delta receive signal detect
2	Parity error	2	Trailing edge ring detector
1	Overrun error	1	Delta data set ready
0	Data ready	0	Delta clear to send

"Delta" bits indicate a change in the indicated flags since the last read.

The service to be called is placed into AH. Parameters for the service are placed in AL. The interrupt is called, and the results placed in AL (or AX for service 3).

```
mov   AH, 0
int   14H
```

2.6.12 Serial port operation in BASIC

Serial communications can be easily implemented in BASIC. This language provides statements which allow programming of the UART without reference to the actual I/O port memory addresses. The serial port initialisation parameters are set using the OPEN statement:

```
OPEN "COM1:9600,N,8,1" AS #1
```

which initialises COM1 at 9600 baud, no parity, 8 data bits, 1 stop bit. Data is written to or read from a "file" numbered "1".

To read 1 byte from COM1, we write: `A$=INPUT$(1,#1)`

The byte is read from the receive buffer in the UART and converted to an ASCII character and then assigned to a string variable A$. To display the decimal number actually read, we can use the ASC function:

```
PRINT ASC(A$)
```

> The INPUT$ function is the preferred method of reading data from the serial port. Other statements such as INPUT and LINE INPUT may work, but may give unpredictable results if the data in the input stream contains ASCII control characters such as LF and CR. If we were to use INPUT, then the input would stop when the incoming data contained a comma or a CR character. This is OK for reading in data from the keyboard, but not from a file where we may wish to capture all the data.

In Visual Basic (VB), the procedure is very similar. The COMM port object is placed on a form, in the example below, the form is called d frm_MainMenu and the COMM port object is called MSComm1.

MSComm1 has properties that can be set in code that allow the serial port to which is assigned to be configured.

```
frm_MainMenu.MSComm1.Settings = "9600,E,7,1"
frm_MainMenu.MSComm1.InputLen = 0
frm_MainMenu.MSComm1.RTSEnable = True
frm_MainMenu.MSComm1.DTREnable = False
frm_MainMenu.MSComm1.PortOpen = True
```

Methods available to the serial port object allow characters to be read and assigned to a variable, or the value of a variable to be written and transmitted from the computer:

```
ReadChar = frm_MainMenu.MSComm1.Input
frm_MainMenu.MSComm1.Output = WriteString + vbCr
```

2.6.13 Hardware handshaking

Although at first sight reading the serial port using BASIC appears fairly
straightforward, difficulties arise when the data cannot be read fast enough
and the input buffer overflows. The buffer holds 255 characters (i.e.
1 character = 1 byte). **Handshaking** (either software codes or hardware
signals) is used to halt transmission of data from the sending computer until
the receiving computer has emptied the buffer. Various functions are
available in BASIC to allow either software or hardware handshaking.

- LOC(x) returns the number of characters in the input buffer for file number "x".
- LOF(x) returns the number of character spaces available in the input buffer.
- EOF(x) returns (−1) if the buffer is empty or 0 if it is full.

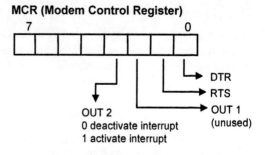

MCR (Modem Control Register)

OUT 2
0 deactivate interrupt
1 activate interrupt

DTR
RTS
OUT 1
(unused)

For COM2, the MCR port address is 2FC

For the serial data acquisition circuit, hardware handshaking must be used
since there is no method of interpreting software codes such as XON and
XOFF. The RTS line offers the most convenient form of hardware
handshaking. RTS is arranged to go logic high (−10 V on RS232 signal
lines) when the buffer is full, and then low when the buffer is empty. Now,
the RTS signal line is available via the 2nd bit in the Modem Control
Register. Setting this bit to 0 will actually set RTS to logic high. The
BASIC OUT statement can be used to send a byte to an I/O port address.
Thus:

    OUT &H2FC, &H8   Set RTS logic high

    OUT &H2FC, &HA   Set RTS logic low

These statements, in combination with the LOC, LOF and EOF functions, can be used to control RTS. The RTS signal in turn can be wired to halt and resume transmission of data as required.

Note: The BASIC INPUT$ statement makes use of
interrupts to read the data from the serial port. Make
sure that OUT2 remains set at 1 when writing data to
the MCR.

2.6.14 RS485

The maximum distance allowed by RS232 is about 15 m which in an industrial environment can be a severe limitation, especially when the computer is located say in a control room some distance away from the transducer. Further, the maximum **data transfer rate** can be a limitation for fast data acquisition. Standards such as RS422 and **RS485** were developed to overcome these limitations and permit greater flexibility and performance for instrumentation applications.

An increase in transmission speed and maximum cable length is done by using **voltage differentials** on signal lines A and B. For a space, or logic 0, the voltage level on line A is greater than that on line B by 5 V. For a mark, or logic 1, the voltage level on line B is greater than that on line A. The receiver inputs on the line driver chip determine whether or not the signal is mark or space by examining the voltage difference between lines A and B. Two signal wires are thus required for data transmission.

Unlike RS232, in which there is usually a connection between two pieces of equipment, the RS485 standard allows for up to 32 line drivers and 32 receivers on the one set of signal lines. This is achieved by **tri-state** logic on the line driver pins.

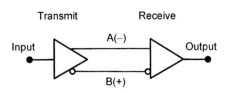

In tri-state logic, pins can be at logic 0, logic 1 or high impedance. The last state effectively disconnects the driver from the line. The high impedance state is set by an "enable" signal on the driver chip.

Features of RS485:

- Maximum distance 1200 m.
- Data rate up to 10 Mbps.
- 32 line drivers and receivers on the same line.
- TTL voltage levels.

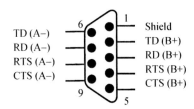

Common RS485 9-pin connector

2.6.15 GPIB

The General Purpose Interface Bus (**GPIB**) or **IEEE 488** interface was developed in 1965 by Hewlett Packard for connecting multiple scientific measuring instruments together. Up to 15 devices may be attached to the interface lines (or **bus**). One of the devices can be activated as the '**controller**'. Control can be passed to another device if required.

The interface consists of:

- 8 data lines
- 3 data control lines
- 5 management lines
- 8 ground lines

GPIB connector

8-bit data can be transmitted in parallel, each bi-directional line carrying 1 Mbit/second. A maximum total cable length of 20 m with a maximum separation of 4 m between devices is recommended. Bus extenders and expanders can also be used.

Data
Control
Handshake

7
0
IFC
ATN
SRQ
REN
EOI

DAV
NRFD
NDAC

Devices may be configured as listeners, talkers and controllers.

- **Listener**: A device set to be a listener can accept data over the bus from another device. More than one listener at any one time is permitted. A listener receives data when the controller signals it to read the bus.
- **Talker**: A device set to be a talker can send data to another device on the bus. Only one talker can be specified at any one time. The talker waits for a signal from the controller and then places its data on the bus.
- **Controller**: A controller can set other devices as listeners or talkers or to take control. The presence of a controller is optional. For example, if there is only one talker and all other connected instruments are listeners, then no controller is required.

Data is put onto the data bus by a talker when no device has pulled NRFD (not ready for data) low (negative logic). DAV (data valid) indicates that data is ready, and all devices then pull NDAC (no data accepted) low until the data is read. The **parallel connection** of devices ensures that NDAC goes high again when all listeners have accepted the data.

Each device responds to commands sent over the data bus. Each device can recognise its address when it appears on the data bus. The device address is usually set by dip switches or from software. The management of the bus is done by the controller which typically contains a special purpose IC on a GPIB **interface card**.

2.6.16 USB

The range of peripheral devices now connected to personal computers are attached by serial, parallel and PS/2 ports, and the requirement for ease of use has resulted in the development of the **Universal Serial Bus (USB)**. USB is designed to be a low cost, expandable, high speed, serial interface that offers "**plug and play**" functionality primarily for business and consumer peripherals.

Data transfer rates for the first implementation of USB were up to 12 Mbps. USB 2.0 allows up to 480 Mbps making it suitable for real time video and audio, high resolution digital cameras and data storage devices.

A USB system consists of:

- USB interconnect
- USB devices
- USB host controller

The connection between USB devices and the host, the data flow protocols, and the manner in which devices are addressed.

USB devices are either **hubs**, which provide attachments to other hubs or actual devices. The host controller queries the hubs to detect the attachment or removal of devices. A unique **bus address** is assigned by the host when a device is connected.

There is only one host in any USB system. The host controller sends a **token packet** that describes the type and direction of the data, the device address, and the endpoint number. The device that is addressed then receives a **data packet** and responds with a **handshake packet**.

The USB transfers signal and power over a four-wire cable. Differential signalling occurs over two of the wires. There are three data rates:

- high speed at 480 Mbps
- full speed at 12 Mbps
- low speed at 1.5 Mbps

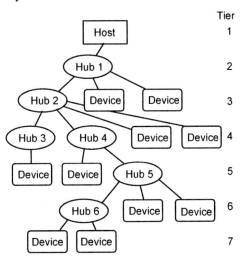

A maximum of seven tiers is allowed. Tier 1 is the root hub and Tier 7 can only contain devices.

In the USB system, one device must be the host and this places some restrictions on its use in an industrial setting. A simple modem, for example, can be wired using a **null modem** connection and be used with a PLC or other RS232 supported transducer. With a USB system, a computer must generally act as a host, even if communication is wanted only from one device to another.

The '**on the go**' (**OTG**) supplement to the USB 2.0 standard allows some degree of peer to peer communication without the need for a fully featured host.

In RS232 communications, the format of the data is not defined – it is usually ASCII text but need not be so. The USB uses layers of transmission protocols to transmit and receive data in a series of **packets**.
Each USB transmission consists of:

- **token packet**
- optional **data packet**
- **status packet**

The USB host initiates all transactions. The token packet describes the type of communication (read or write and the destination address). The data packet contains the data to be communicated. The handshaking packet reports if the data was received or transmitted successfully.

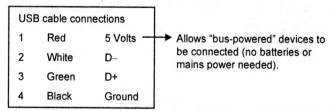

USB is designed to be '**plug and play**'. When a device is plugged into the bus, the host detects its presence by signal levels on the data lines. It then interrogates the new device for its device descriptor, assigns a bus address to the device, and then automatically loads the required device driver. When the device is unplugged, the host detects this and unloads the driver. This process is called **enumeration**.

> **Firewire (IEEE 1394)** is a serial interface standard originally developed by Apple Computer (ISB was originally developed by Intel). Firewire allows up to 400 Mbps and is a competitor to ISB (when first introduced, Firewire was several times faster than the then USB standard). Like USB 2.0, the main consumer benefit is high speed for video capture from digital cameras and camcorders without the need for dedicated video capture interface cards.

2.6.17 TCP/IP

TCP stands for Transmission Control Protocol, and IP stands for Internet Protocol. **TCP/IP** is a set of protocols that allows computers to communicate over a wide range of different physical network connections. TCP/IP provides protocols at two different layers of the **OSI Reference Model**. In everyday terms, the **world wide web** (www) and **email** (SMTP) make use of TCP/IP to communicate over the internet which in turn runs on a variety of packet switching network systems the most popular of which is the **Ethernet**. The actual connections between host computers is done by satellite, coaxial cable, phone lines etc. For interfacing applications, the internet is useful for communicating commands and results from a remote sensor, but would be unsuitable for a direct interface to the transducer due to the response time of the process.

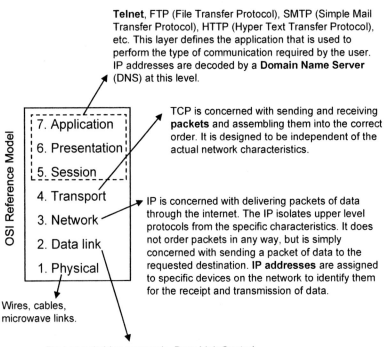

Telnet, FTP (File Transfer Protocol), SMTP (Simple Mail Transfer Protocol), HTTP (Hyper Text Transfer Protocol), etc. This layer defines the application that is used to perform the type of communication required by the user. IP addresses are decoded by a **Domain Name Server** (DNS) at this level.

TCP is concerned with sending and receiving **packets** and assembling them into the correct order. It is designed to be independent of the actual network characteristics.

IP is concerned with delivering packets of data through the internet. The IP isolates upper level protocols from the specific characteristics. It does not order packets in any way, but is simply concerned with sending a packet of data to the requested destination. **IP addresses** are assigned to specific devices on the network to identify them for the receipt and transmission of data.

OSI Reference Model

7. Application
6. Presentation
5. Session
4. Transport
3. Network
2. Data link
1. Physical

Wires, cables, microwave links.

Packet switching protocols. Data Link Control (DLC), **Ethernet**. This layer is concerned with the transmission of packets in a specific mode for delivery over the mechanism of the physical layer. Characteristics of the Ethernet determine the response time of the network.

2.6.18 Review questions

1. What basic digital logic circuit forms the basis of a UART used for serial communications?

2. The RS232 serial interface is very popular because of its availability and simplicity. Without actually describing the sequence of events, describe the function of the six most commonly used signal lines.

3. What is a 'null modem' and under what circumstances should it be used?

4. In a UART, why would the receiver clock be made to oscillate 16 times the rate of data transmission?

5. What is a parity bit?

6. Why is a 232 line driver IC required for communications between two UARTs which take place over an external signal cable?

7. Why, in serial communications, is there a choice of 5, 6, 7 or 8 data bits for the data being transmitted or received? (Hint: Investigate the ASCII code for character transmission.)

8. What is the main advantage of the RS485 interface standard?

9. What advantage would a GPIB interface have over an RS232 communication in an interfacing application?

2.7 Programmable logic controllers

2.7.1 Programmable logic controllers

Industrial processes have been traditionally controlled by electromechanical systems (i.e. switches and relays). An electromechanical control system controls the output states according to the input states. The logic behind the switching and the resultant actions depend upon way the switches and relays are wired together. The overall function in these systems is not so easily changed and the systems are not easily maintained.

In the late 1960s, digital controllers were introduced to allow some degree of programming to control the sequence of operations required in an industrial process. A **programmable logic controller** (PLC) examines its input states and turns off or on its outputs according to a pre-loaded program that can be easily altered to suit changed circumstances. Advances in technology have resulted in programmable controllers that can communicate with each other, as well as receive and transmit control data to remote locations. Present day systems feature functional block diagrams and structured programming in a standardised way.

PLC devices have standard input and output interfaces. Standard input interfaces allow direct connection to process transducers. Standard output interfaces allow direct connection to relays or circuits that energise process actuators.

PLC devices operate under program control. A program consists of a series of statements or logic instructions called a **list**. The control unit **scans** the inputs and performs logical operations according to the loaded program and then switches the outputs accordingly. The inputs are then scanned again and the cycle repeats. The result is an industrial **process**.

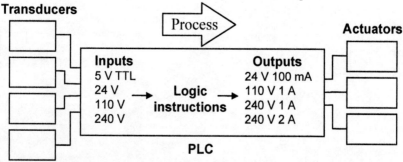

A PLC itself consists of the control unit (or CPU) as well as connections for input and outputs, RAM memory (≈ 10 Kb) and a power supply. Inputs are opto-coupled to the input circuits in the PLC to protect the PLC from noise.

2.7.2 Timing

A PLC program consists of a series of instructions that represent logical operations performed on the inputs. The state of the outputs is set or cleared in accordance with the logical result of the program instructions.

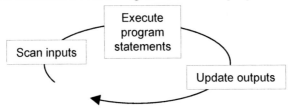

The states of all the inputs are copied into RAM before the program instructions are executed. Instructions are then processed in sequence. The resulting output states are stored in RAM. When the program execution is completed, the stored output states are transferred to the output terminals of the PLC.

Since it takes a finite time for the PLC to read the inputs (**input response time**), process the instructions (**execution time**) and set the outputs (**output response time**), changes to the inputs can only be registered if they last longer than the **scan time** (input response time + execution time + output response time). If the input changes more rapidly than this, then the PLC may not detect the change and the required output may not be set. This is called a **phasing error**.

When an input changes state, the resulting change in the output will take place, in the worst case, two scan times (less one **input response time**) later.

PLC scan times are usually quoted in terms of the length of time to execute a 1024-byte program and depend upon the clock rate of the controller. **Scan times** are usually in the order of 5 to 10 msec.

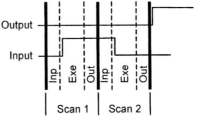

In some PLCs, the program is executed line by line. The control unit scans the inputs actually referenced by each program statement as it is executed. The output named in each program statement is then updated according to the logical operation in the program statement and then held or latched in that state.

2.7.3 Functional components

The PLC inputs consist of input relays or **contacts** which may be physically real devices or simulated as labelled contacts in the program. The outputs are called **coils** representing the coil of a relay. The outputs can take the form of transistor switches, triacs or relays. As well as inputs and output devices, the control unit also typically contains latches, counters, timers and registers for data storage. Some PLC devices also have analog I/O capability.

Latch	When set to true, the output of the latch will stay on until the latch is reset.
Counter	Counts pulses at its input and sets the timer output to true when a preset number of counts have been registered. True at the reset input resets the accumulated count to zero.
Timer	Output of the timer turns on or is true a preset number of seconds after the input is true. When the input is set to false, the timer is reset and the output is set to false. While the timer is counting down, the PLC continues to scan and execute its instructions. Input and output errors can occur with timers since the input to the timer may not be registered until it is scanned. Further, the output device may not be energised until the PLC has completed executing the program.
Registers	Used for storing data or the results of logical operations as bits (true 1 or false 0). Registers are similar to internal contacts. The contents of registers may be shifted left or right. Data bits can be moved into and out of registers using a MOV instruction.

A PLC can generally communicate with external devices using RS232 serial communications. The PLC may be fitted with an RS232 serial port for this purpose. Data may be sent to or received from the PLC. The data can be stored in the PLC data memory.

The actual logic program run by the PLC is usually entered through a keypad or downloaded from a microcomputer. Before a program is executed, various levels of verification are performed to ensure that the program was transferred successfully into the CPU of the PLC. Various systems are employed to protect processes and plant in the event of a power loss. PLC systems are designed to be robust and operate unattended in an industrial environment.

2.7.4 Programming

PLCs act upon **list code** instructions. To facilitate the creation of a list code, **ladder logic diagrams** are used to simulate the existence and actions of input and output devices. A ladder diagram consists of two vertical rails inside which are placed symbols for contacts, relays, functions, and logical operations on **rungs**. Logic flow (or current flow) is from left to right and top to bottom on the diagram. An output on a rung is energised if there is a continuous path of true logic leading back to the left (or positive) rail of the ladder (**logic continuity**).

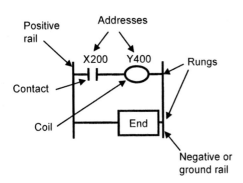

Each rung must contain one or more inputs and one or more outputs. The first object on a rung must be an input and the last object on a rung should be an output, a counter, timer or an internal relay. The last rung in a ladder diagram is an END instruction.

Basic instructions

Symbol	List code	Action
⊣⊢	LD	Evaluates to true if the physical contact it represents is closed or on.
⊣/⊢	LDB	Opposite to LD.
─O─	OUT	Evaluates to true, and thus energises a normally open coil, if there is a continuous "true" path from the left-hand side of the ladder to it.
─⊘─	OUTB	Opposite to OUT

The PLC scans the ladder diagram from top to bottom and left to right. In a load instruction (LD) the physical state of a scanned input is determined and the symbol in the ladder diagram evaluates to true if the physical device is closed or on. The symbol may also be used for internal utility relays or switches that do not physically exist.

2.7.5 Ladder logic diagrams

Ladder diagrams can become quite complex. PLC systems generally have the ability to perform math functions on data, apply Boolean operators, and store data in registers or memory locations.

Consider these simple examples:

An output is only energised when there is a continuous true path from the left-hand side to the right-hand side of the ladder.

Ladder diagram	List code

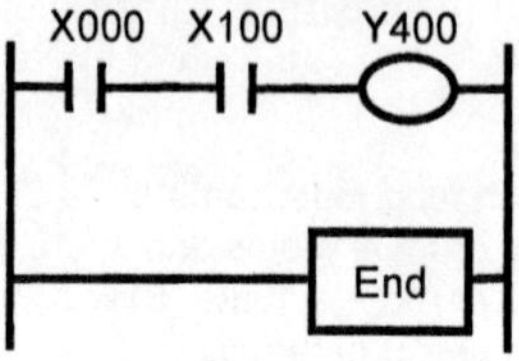

LD X000
AND X100
OUT Y400
END

Output Y400 is energised (true) as long as inputs X000 AND X100 are both closed (true).

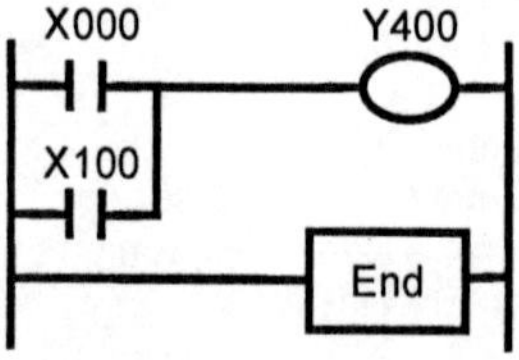

LD X000
OR X100
OUT Y400
END

Output Y400 is energised (true) as long as input X000 is closed OR X100 is closed (true).

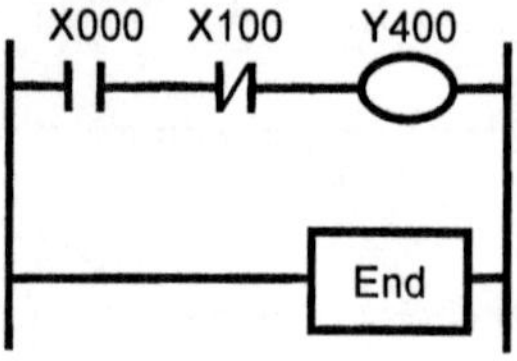

LD X000
ANI X100
OUT Y400
END

Output Y400 is energised (true) as long as input X000 is closed AND X100 is open (false).

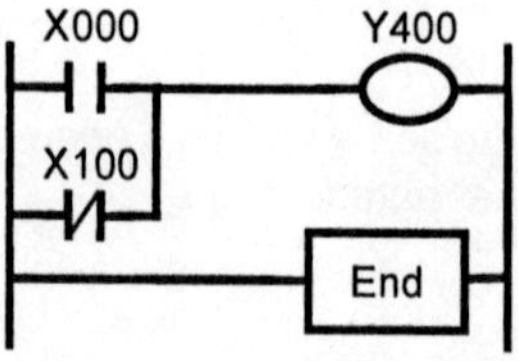

LD X000
ORI X100
OUT Y400
END

Output Y400 is energised (true) as long as input X000 is closed OR X100 is open (false).

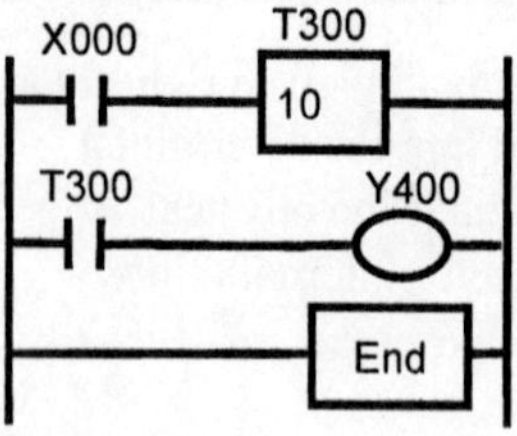

LD X000
OUT T300
K 10
LD T300
OUT Y400
END

When X000 turns on (true) then timer T300 begins counting down. After 10 seconds, switch contacts for the timer T300 are closed (true) and output Y400 is energised (true).

The example below shows a timer circuit whose output device turns on and remains on for the time period when an input pulse appears at the input. This circuit uses an **internal relay**. Internal relays are coils and contacts that are simulated by the PLC in memory. Like external relays, they consist of an output coil and a set of contacts that can be used as the input to other objects on a ladder rung.

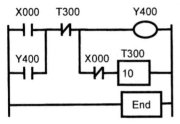

In this example, when the input X000 is true, there is logic continuity through the normally closed timer contacts T300 to the output Y400. This true state is fed back into the input to the normally closed contacts of the timer. Thus, when the input X000 goes false, the output Y400 remains on, it is **latched** by its own contacts.

Now, when X000 goes false, a true signal is sent to the timer to begin the countdown period. During the countdown period, the output device Y400 remains energised by the latched path through the contacts Y400. When the countdown period has expired, the normally closed contacts of T300 become open, thus interrupting the logic continuity to the output Y400 and so Y400 is de-energised. The ladder logic above acts like a **pulse extender**. A short pulse on the input X000 can be extended into a longer pulse appearing at the contacts of the output Y400.

Ladder logic diagrams can easily become unwieldy and difficult to maintain unless a certain methodology is followed to give them structure.

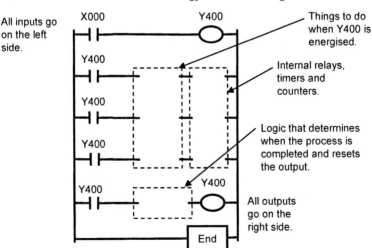

2.7.6 PLC specifications

There are numerous PLC systems available. The table below gives typical specifications and features.

- 16 optically isolated AC inputs, 100 V–220 V, 50 Hz to 60 Hz.
- 8 optically isolated DC inputs, 5 V–24 V.
- 8 analog inputs, 0–5 V.
- 2 high speed optically isolated inputs up to 100 KHz.
- 8 AC optically isolated outputs, up to 220 V, 8 A.
- 8 DC optically isolated outputs, 48 V, 8 A.
- 32 timers, 16 on 0.01 second base, 16 on 1 second base.
- 32 up or down counters.
- 256 internal relays.
- Real time clock with time and date.
- RAM back-up.
- Data logging for 16k bytes.
- Programmable 4.5 digit LED display.
- Internal 110 V/220 V AC power supply.
- Ladder logic in EEPROM for 2 programs of 4k 32-bit words each.
- Scan rate 7 msec/k 32-bit instructions.
- Real time scan rate indicator.
- Network and I/O expandable via RS485.
- Heavy duty anodized aluminium alloy construction.
- EMI immunity and efficient heat dissipation.
- Standard DIN 43700 instrument case size.

PLC control panel for chilled water supply

2.7.7 Review questions

1. The scanning operation that is the feature of a PLC can be considered to be similar to the multitasking mode of operation of an event-based applications language like Visual Basic. Would you agree?

2. Design a ladder logic diagram that switches on a refrigerator compressor motor when the temperature rises above a preset limit and switches it off when the temperature falls below another preset limit.

3. Design a ladder logic diagram that will control a pedestrian crossing with a set of traffic lights. A press button on each side of the street act as inputs. "Red", "Amber" and "Green" traffic lights, "Walk" and "Don't Walk" indicator lights are the outputs. When a pedestrian presses a button momentarily, there is a 60 second delay before the "Green" lights are extinguished and the "Amber" lights are illuminated. "Amber" is illuminated for 10 seconds, then is extinguished, and "Red" is illuminated. There is now a 2 second delay before the "Walk" signs are illuminated. "Walk" is illuminated for 60 seconds after which time it is extinguished and the "Don't Walk" signs are flashed on and off for 10 seconds. After 10 seconds, "Don't Walk" is held continuously on. At this time, there is a 2 second delay before the Red lights are extinguished and the Green lights are illuminated. The Green lights are kept illuminated if there is no momentary press of the pedestrian button. If a pedestrian presses a button more than once while the Green light then the system only responds to the first press. If a pedestrian presses a button during the "Walk" and flashing "Don't Walk" parts of the cycle, then these presses are ignored.

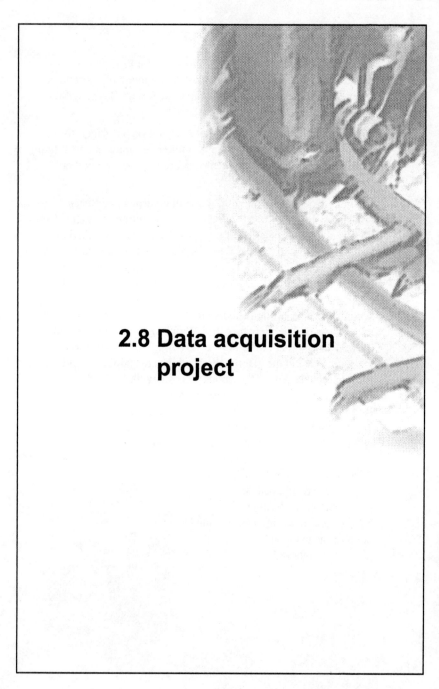

2.8 Data acquisition project

2.8.1 Serial data acquisition system

This project is an analog to digital data acquisition system that reads temperature from a thermocouple and interfaces this to the computer's serial port. No special computer interface card is required. This type of interface is suitable for relatively slowly changing physical quantities. The interface system requires a ±5 V power supply and a modest number of integrated circuits which are readily available from electronics parts suppliers. It can be used with virtually any PC equipped with a serial port.

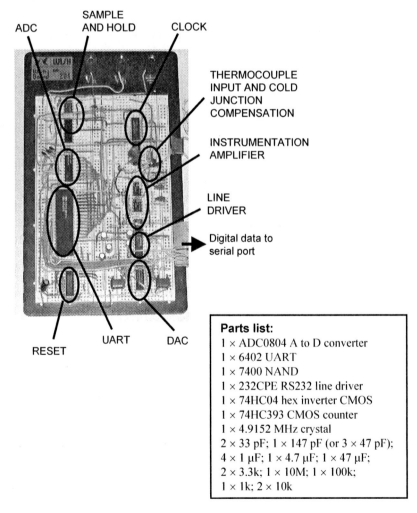

SAMPLE
ADC   AND HOLD   CLOCK

THERMOCOUPLE
INPUT AND COLD
JUNCTION
COMPENSATION

INSTRUMENTATION
AMPLIFIER

LINE
DRIVER

Digital data to
serial port

RESET   UART   DAC

Parts list:
1 × ADC0804 A to D converter
1 × 6402 UART
1 × 7400 NAND
1 × 232CPE RS232 line driver
1 × 74HC04 hex inverter CMOS
1 × 74HC393 CMOS counter
1 × 4.9152 MHz crystal
2 × 33 pF; 1 × 147 pF (or 3 × 47 pF);
4 × 1 μF; 1 × 4.7 μF; 1 × 47 μF;
2 × 3.3k; 1 × 10M; 1 × 100k;
1 × 1k; 2 × 10k

The ADC converts a 0–5 V analog signal to an 8-bit digital value (0–255). This digital value is passed through a UART which serialises the data into a bit stream for transmission over a single wire to the computer's serial port.

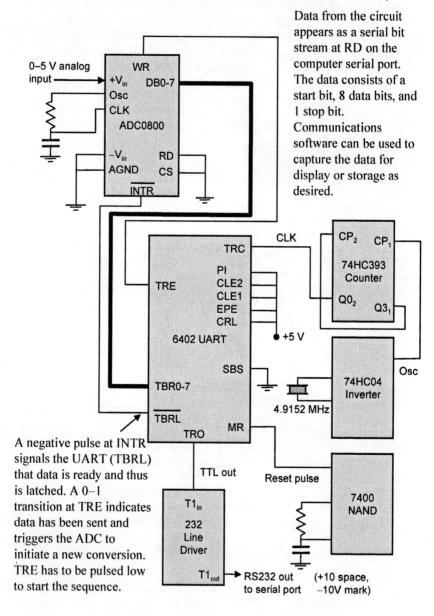

Data from the circuit appears as a serial bit stream at RD on the computer serial port. The data consists of a start bit, 8 data bits, and 1 stop bit. Communications software can be used to capture the data for display or storage as desired.

A negative pulse at INTR signals the UART (TBRL) that data is ready and thus is latched. A 0–1 transition at TRE indicates data has been sent and triggers the ADC to initiate a new conversion. TRE has to be pulsed low to start the sequence.

2.8.2 Circuit construction

1. Connect an ADC0804 IC as shown in the figure below. Use +5 V for V_{cc}.

2. Connect 2 V DC to $+V_{in}$ (analog input) and measure the voltages that appear on the digital outputs.

3. Comment on the binary number indicated by these voltages (DB7 to DB0).

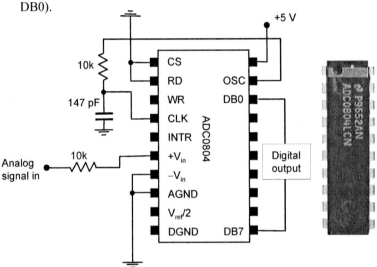

4. Position the UART on the laboratory breadboard and configure the device for 8 data bits, no parity, 1 stop bit.

5. Set CRL, PI and EPE to 1, and SBS, SFD, and RRD to 0. Use +5 V as V_{dd}.

6. Now, we wish to indicate to the UART that data appearing at TBR1 to TBR8 is valid and ready to transmit when the ADC conversion is completed. Select a suitable signal line from the ADC and connect to $\overline{TBRL}$ on the UART.

7. We also wish to initiate a new conversion at the ADC after the data at the UART has been sent. Select a suitable signal line on the UART and connect to WR on the ADC.

Character length select	CLS1	CLS2
5	0	0
6	1	0
7	0	1
8	1	1

Parity	PI	EPE
None	1	x
Even	0	1
Odd	0	0

SBS, stop bits, 0 for one, 1 for two

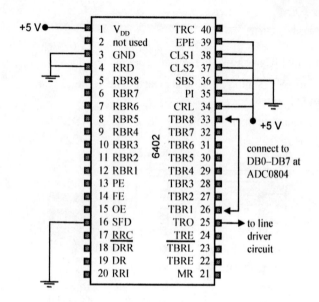

8. Place the line driver 232CPE chip on the laboratory breadboard.

9. Connect the TTL output from the UART to a TTL input on the 232CPE chip. Note: the 232 chip is a dual IC with two sets of separate drivers. Select either T1 and R1 or T2 and R2.

10. Connect capacitors, resistors and supply voltage to the 232CPE chip as required. Note polarity of the capacitors.

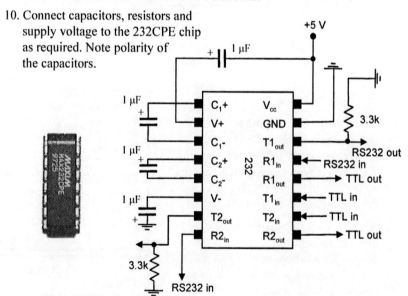

11. Construct a crystal oscillator using high speed CMOS inverters 74HC04.

12. Divide the clock signal down to obtain a baud rate of 9600 using a high speed CMOS counter 74HC393.

13. Connect the stepped-down clock signal to the transmitter clock input (TRC) on the UART.

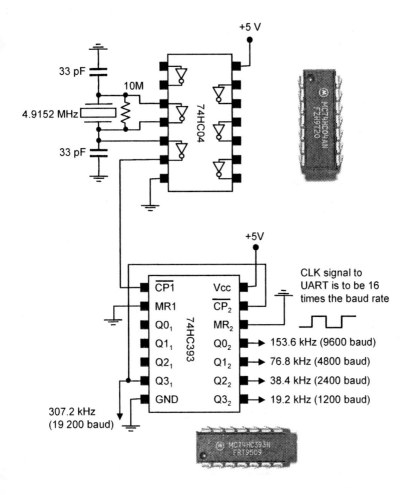

14. Construct a master reset circuit for the UART using a 7400 NAND gate and associated components. Calculate suitable values of R_1 and C to give a time constant of about 2 sec.

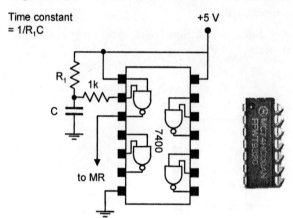

Time constant
= $1/R_1C$

15. Connect the output of the master reset circuit to MR on the UART.

16. The interface circuit is now ready for testing. The first step in testing the circuit is to determine whether or not there is a clock signal at the UART. Display the signal at TRC on a CRO and rectify any wiring errors in the clock circuit.

17. Next, tie the analog input to ground (0 V) and check the voltages at the digital output of the ADC. They should all be at 0 V. Rectify any wiring errors before proceeding.

18. Check the TTL signal at the output of the UART (TRE). There should be a +5 V pulse (the stop bit) after a series of lows (0 V) which is the data.

19. Check the signal at the output of the line driver to ensure that these same logic levels are represented in −10 to +10 logic.

20. Now apply a small DC signal to the analog input on the ADC and verify that the output from the 232 line driver is consistent with that applied. (The magnitude of the byte appearing as data in the bit stream on the output of the line driver should be consistent with the magnitude of the DC input signal from 0–5 V range.)

21. Configure a 25 pin or 9 pin connector cable as a null modem, with hardware handshaking (connect RTS on the computer serial port plug to one of the spare RS232 inputs on the line driver IC).

22. Now, some handshaking is required. When the receive buffer is full, RTS is set high by the receiving computer or our applications program. When the receive buffer is empty, RTS is set low. When the transmit buffer on UART is empty, UART sets TRE high. When transmission is in progress, TRE is low. We wish to arrange things so that when RTS is low and TRE goes from low to high, the WR signal on the ADC goes from low to high and remains high during conversion.

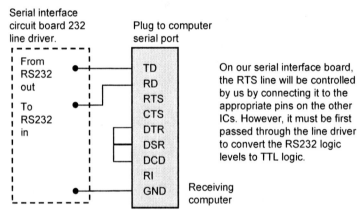

Serial interface circuit board 232 line driver.

Plug to computer serial port

From RS232 out

To RS232 in

TD
RD
RTS
CTS
DTR
DSR
DCD
RI
GND

On our serial interface board, the RTS line will be controlled by us by connecting it to the appropriate pins on the other ICs. However, it must be first passed through the line driver to convert the RS232 logic levels to TTL logic.

Receiving computer

When the receive buffer is full, RTS is set high by the receiving computer (−10 V on RS232 signal lines from line driver). When the receive buffer is empty, RTS is set low (+10 V on signal lines). When the transmit buffer on the UART is empty, the UART sets TRE high. When transmission is in progress, TRE is low. We wish to arrange things so that when RTS is low and TRE goes from low to high, the WR signal on the ADC goes from low to high initiating a new conversion WR and remains high during conversion.

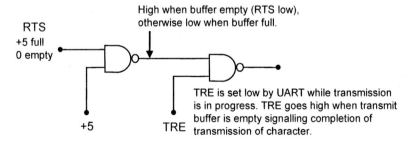

High when buffer empty (RTS low), otherwise low when buffer full.

RTS
+5 full
0 empty

+5

TRE

TRE is set low by UART while transmission is in progress. TRE goes high when transmit buffer is empty signalling completion of transmission of character.

Consider the logic circuit on the previous page. When transmission is in
progress, and the buffer is empty or full, output is high. When the
transmission is complete, the output is low if the buffer is empty, otherwise
high. Thus, if this output goes low, transmission is complete and receive
buffer is empty, therefore a new conversion should be initiated. However, a
new conversion requires a low to high transition on WR. Hence, if this
signal is inverted, then the required action is obtained. That is, when
transmission is complete and buffer is empty, final output goes from low to
high and remains high. A new conversion is initiated, which when
complete, INTR signals the UART to transmit. TheUART sends TRE low
while transmission is in progress which sends WR low.

An inverter can be fashioned from
a NAND gate as follows:

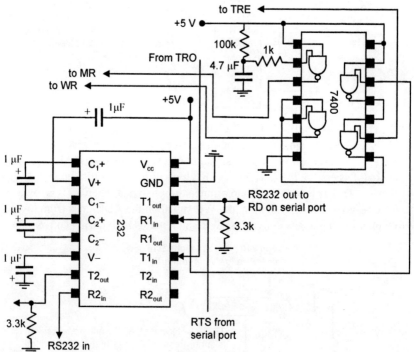

2.8.3 Programming

In this part of the project, a computer program is required to operate the serial data acquisition system. The program is to operate the control lines and data signal line of the serial port. Such a program can be implemented in low level assembly language or an applications language like BASIC. The procedure here assumes that the serial data acquisition system has been built and is working properly. This can be verified using an oscilloscope. With RTS from the computer's serial port held low (+10 V on actual signal line), then there should be a bit stream of data at RD.

Assembly language

1. In this part of the project, an assembly language program will be written to perform the following steps:
   (a) Initialises the serial port (COM1 or COM2).
   (b) Causes RTS to be set at logic high (−10 V on RS232 signal line).
   (c) Reads in one byte from the serial port.
   (d) Displays the value on the screen.
   (e) Causes RTS to be set logic low (+10 V).
   (f) Allows the program to be terminated by pressing any key on the keyboard.

To initialise the serial port, BIOS service routine 0 is used. The port to be initialised (0 for COM1, 1 for COM2) is specified in the DX register. The initialisation parameters are assembled into a byte from the information given in the table.

The service to be called (0) is placed into AH. The byte containing the initialisation information is placed in AL. The serial port service is called through interrupt 14H. Thus, to initialise the serial port COM2 to 9600 baud, 8 data bits, 1 stop bit and no parity, the following assembly language instructions are required:

```
MOV DX,01H      ;SELECT COM2
MOV AH,0        ;INITIALISE SERIAL PORT
MOV AL,0E3H     ;9600,N,1,8
INT 14H
```

There are four BIOS services available for the serial port. The number for the service to be called is placed into AH. Parameters or data for the service are placed in AL. Interrupt 14H is called, and any results placed in AL (or AX for service 3).

BIOS services

0	initialise serial port
1	send one character
2	receive one character
3	read serial port status

Bits 7 6 5	Baud Rate	Bits 4 3	Parity	Bit 2	No. stop bits
0 0 0	110	0 0	none	0	one
0 0 1	150	0 1	Odd	1	two
0 1 0	300	1 0	None		
0 1 1	600	1 1	Odd	**Bit** **1 0**	**Data** **bits**
1 0 0	1200			0 0	unused
1 0 1	2400			0 1	unused
1 1 0	4800			1 0	7
1 1 1	9600			1 1	8

Despite these services being available, they are actually particularly unhelpful as we shall see.

2. Now, the serial interface board is controlled by a signal on the computer's RTS line. To start analog to digital conversions, the voltage on the RTS line needs to be set at –10 V (which is TTL logic high). This is called hardware handshaking. There is no BIOS service available which controls RTS and so writing directly to the modem control register (MCR) will be necessary. To start conversions, a logic 1 is placed in the RTS bit of the MCR (and OUT2 is also set to 1 to allow the BIOS routines to work).

Determine a bit pattern and hence a hex number to write to the MCR to enable conversions. Then determine the hex number to write to the MCR to inhibit conversions.

Note: A logic 1 in the RTS bit sets the RTS logic level low (+10 V on signal line) which starts conversions on the interface card.

MCR (Modem Control Register)

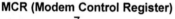

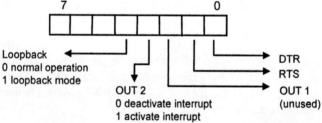

Loopback
0 normal operation
1 loopback mode

OUT 2
0 deactivate interrupt
1 activate interrupt

DTR
RTS
OUT 1
(unused)

Purpose	COM1	COM2
Tx,Rx data	3F8	2F8
Interrupt enable	3F9	2F9
Interrupt ident	3FA	2FA
Line control	3FB	2FB
Modem control	3FC	2FC
Line status	3FD	2FD
Modem status	3FE	2FE

3. The hex codes which set and clear the RTS line need to be written to the MCR which, for COM2, is located at I/O port address 2FCH. The MOV instruction cannot be used for this since MOV writes data to either registers or regular memory locations. The assembly language instructions IN and OUT are used to write to port addresses.

4. To read in a byte from the serial port, the BIOS "receive one character" service and call interrupt 14H can be used. However, when this is used, interrupt 14H clears the MCR and control of RTS is lost. Thus, RX/TX register needs to be read (located at the I/O port base address 3F8H or 2F8H) directly with the IN instruction.

Now, the OUT instruction has two forms. The first form is:

```
OUT d8, AL or AX
```

d8 is an 8-bit port address from 0 to 255 which is fine if a write to a port with an address in this range is required. If a write to a port with a higher port address is needed, then this 16-bit number needs to be loaded into DX and then the OUT statement is used to obtain the address from DX. Thus, the second form of the statement is:

```
OUT DX, AL or AX
```

A byte or a word is written depending on whether AL or AX is specified as the source. To write a byte to location 2FCH, the byte is loaded into AL, and the number 2FCH is loaded into DX and thus:

```
MOV DX,02FCH
MOV AL,0AH
OUT DX,AL
```

The IN instruction has a very similar syntax:

```
IN AL or AX, d8/DX
```

The source is specified by either the I/O port address in DX or an 8-bit number directly (for ports 0–255). The byte read is written to AL. If AX is specified as the destination, then a word is read from the port. In this case, the port only gives us a byte to read (i.e. I/O port address 3F8H or 2F8H contains 1 byte read from the serial port receive buffer). Thus, AL must be specified as the destination.

5. The serial port can now be initialised (using a BIOS interrupt call),
    conversions can be started and stopped (using OUT to the MCR) and
    data can be read (using IN). It would be convenient to display the byte
    on the screen and also to allow the user to exit the program by pressing
    a key. These last two features can be readily incorporated using BIOS
    service routines.

    Note: The video BIOS services allow a character to be written to the screen. A
    character means an ASCII character. That is, the hex number in AL is treated as
    an ASCII code and the matching character symbol appears on the screen. To
    have the actual number appear, rather than the ASCII interpretation of that
    number, then one has to analyse the number and have the program output the
    ASCII codes corresponding to each digit in the number to be displayed.

    Video services are called through interrupt 10H. To write a character to
    the screen, there are a number of services available. One that can be
    used is service 9H. This service writes an ASCII character to the screen
    at the current cursor position and does not advance the cursor position
    (other services automatically advance the cursor position so that the
    screen would be filled with data in our present application as characters
    were read from the serial port). The character to be written to the
    screen is specified by the number in AL. The service to be executed is
    specified by the number in AH.

    ```
    MOV AH,9H       ;OUTPUT SCREEN SERVICE
    INT 10H         ;BYTE TO OUTPUT IS ALEADY IN AL
    ```

    Keyboard services are called through interrupt 16H. The service to be
    called is specified in AH. We are interested in service 1 which reports
    whether or not there is a character in the keyboard buffer. If there is a
    character in the buffer, then the service sets the zero flag (ZF) to 0. If
    there is no character in the keyboard buffer, ZF is set to 1. A jump
    statement can then be used to terminate the program.

    ```
    MOV AH,01H      ;KEYBOARD SERVICE
    INT 16H
    JNZ END
    JMP READ
    ```

BASIC

1. Write a program in BASIC that will perform the following steps:

   (a) Accept user input which specifies baud rate and COM port.
   (b) Initialises the serial port (COM1 or COM2).
   (c) Causes RTS to be set at logic high (−10 V on RS232 signal line).
   (d) Causes RTS to be set logic low (+10 V).
   (e) Tests to see how many bytes are waiting to be read.
   (f) Tests to see how much space is left in the input buffer.
   (g) Reads in 1 byte from the serial port and displays the value in decimal on the screen.
   (h) Brings all of the above functions together so that the program continuously displays the byte read at the serial port on the screen. At the same time, the program is to halt and continue transmission of data from the serial interface board if the buffer becomes too full or empty as required.

   Note: Make your program user-friendly. That is, allow the program to continuously read the serial port until the user presses Esc at which time the program will close the file handle and exit gracefully to the operating system or the BASIC interpreter prompt.

Questions:
   (a) Calculate the aperture time and maximum frequency obtainable without aperture error for the conversion time you are using in this circuit.
   (b) Calculate the required baud rate that would provide this maximum frequency.
   (c) Compare with the maximum frequency obtained from your circuit.
   (d) Suggest a method (or two) by which the maximum frequency may be increased.

2.8.4 Sample-and-hold

1. Construct a sample-and-hold control circuit using a 7476 JK flip-flop as shown but do not connect R to INTR yet. Rather, supply +5 V to R from the power supply.

2. Test the operation of the flip-flop by providing a negative start pulse and observing the $\overline{Q}$ output. Then put 0 V at the R input and observe the $\overline{Q}$ output.

3. Make sure the flip-flop is working correctly, then disconnect R from the ±5 V supply and connect to INTR on the ADC.

4. Connect the LF398 sample-and-hold IC as shown:

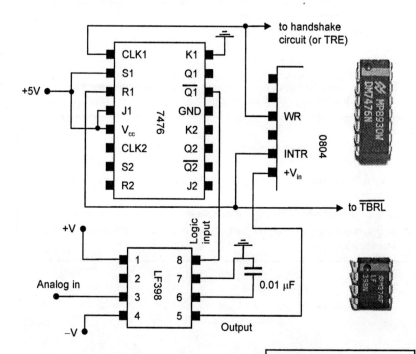

Parts list:
1 × 7476 JK flip-flop
1 × LF398 sample-and-hold
1 × 0.01 μF capacitor

The LF398 requires a bi-polar voltage supply. Ideally, this needs to be a few volts greater than the maximum signal voltage. ± 10 V would be acceptable, but ± 5 V will suffice if a ± 10 V supply is unavailable but the output signal will be reduced in amplitude.

The sequence of operations is:

- Start pulse initiates conversion since it is connected directly to $\overline{WR}$.
- Since $\overline{SET}$ is high, and initially, $\overline{RESET}$ is high, the output $\overline{Q}$ will respond to the clock going low and since J is high (with K low), $\overline{Q}$ is sent low. Sample will be latched on clock signal going low. Conversion actually begins when clock ($\overline{WR}$) goes back high.
- At the end of conversion, $\overline{INTR}$ goes low which sends a low to $\overline{RESET}$ sending $\overline{Q}$ high independent of the signal at J and sending sample and hold to "sample".

5. Using a sine wave signal, compare the waveforms (using a CRO) on the input and the output of the LF398 sample-and-hold IC. Draw these waveforms (you may like to increase the frequency of the signal to about 100 Hz to obtain a clear display). Measure the width (in microseconds) of the steps in the LF398 output.

6. Investigate the signal appearing on the logic input of the LF398 and note the time between each signal pulse.

7. Compare the upper limit to the frequency response of the A to D system with and without the LF398 in use.

Questions:

(a) Make some comment about the width of the steps given by the LF398 IC in relation to the frequency of the input signal.

(b) State whether or not the inclusion of a sample-and-hold circuit improves the maximum frequency which can be digitised without aperture error.

(c) Is there are any other aspect of your system that might cause the upper limit of frequency to be rather low?

2.8.5 Digital to analog system

For the purposes of demonstration, the serial data acquisition system can be easily modified to include a digital to analog facility. Digital data from the computer will be sent (via the line driver IC) to the receive register of the UART. This digital data can then be converted to an analog signal using the DAC0800 IC and monitored on a voltmeter or oscilloscope.

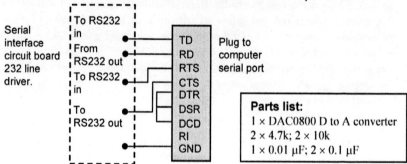

1. The first step in our DAC converter is to obtain a digital signal from the serial port of the computer. Connect the TD signal line from the computer serial port to an available RS232 IN pin on the 232 line driver. Connect the corresponding TTL out to pin RRI (receive register input) on the 6402 UART.

2. Connect CTS from the computer serial port to an available RS232 out. Set CTS high so that the sending computer will send data continuously. Thus, connect the corresponding TTL IN to +5 V on the circuit board.

3. Connect the DAC0800 into the circuit as shown with the digital inputs being connected to the digital outputs from the UART. Also, connect a clock signal to the receiver circuit on the UART by connecting pin 40 to pin 17.

4. Attach a voltmeter to V_o.

5. Modify your interface program to include a write operation. As a demonstration of the DAC operation, the overall procedure is first to feed in an analog signal to the ADC and transmit the digitised output to the computer. Then the computer is going to send the digitised signal back to the circuit board and the DAC is going to convert the signal back into an analog output.

6. With a steady 0–5 V signal applied to the ADC input, and the interface program running, measure the output voltage of the DAC and adjust V_{ref} (if an adjustable power supply is available) so that 255 or FF on the input to the DAC gives +5 V on the analog output. $V_{re} = 2.3$ V should be about right. If no adjustable power supply is available, then set V_{ref} to +5 V and record the analog output voltage for 0 and FF on the digital inputs.

7. Replace the steady DC input analog signal with a sine wave output from a signal generator. Remember, the sine wave input to the ADC has to be positive going always and swing between 0 and 5 V.

8. Attach a CRO to V_o and monitor the analog output signal.

Questions:

(a) Compare the analog output voltage level from the DAC for 0 V and a steady 5 V on the input to the ADC. Can you determine the significance of V_{ref} on the DAC?

(b) Observe the output from the DAC on a CRO when a sine wave is fed into the input of the ADC. Comment on the differences in the shape of the two wave forms.

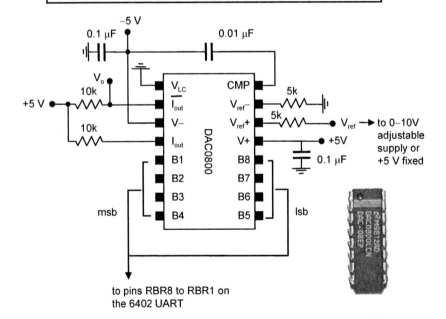

to pins RBR8 to RBR1 on
the 6402 UART

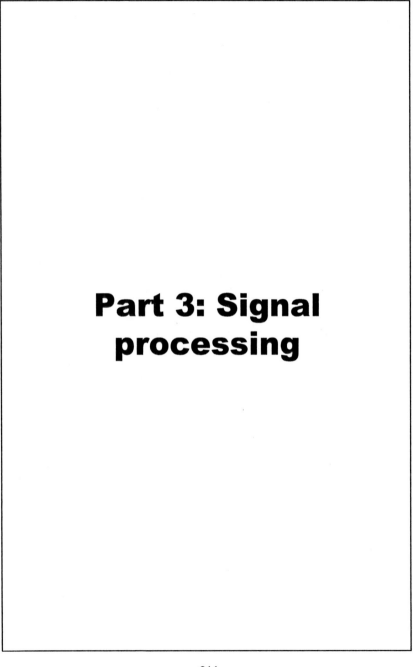

Part 3: Signal processing

3.0 Signal processing

The signals from a transducer are generally of very low magnitude. To prepare them for the computer interface, they must be amplified to an acceptable level and filtered to eliminate unwanted noise. This is the process of **signal processing**.

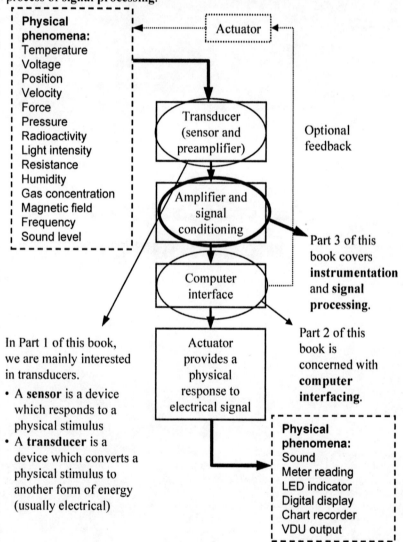

Physical phenomena:
Temperature
Voltage
Position
Velocity
Force
Pressure
Radioactivity
Light intensity
Resistance
Humidity
Gas concentration
Magnetic field
Frequency
Sound level

Actuator

Transducer
(sensor and
preamplifier)

Optional
feedback

Amplifier and
signal
conditioning

Part 3 of this book covers **instrumentation** and **signal processing**.

Computer
interface

In Part 1 of this book, we are mainly interested in transducers.

Actuator
provides a
physical
response to
electrical signal

Part 2 of this book is concerned with **computer interfacing**.

• A **sensor** is a device which responds to a physical stimulus
• A **transducer** is a device which converts a physical stimulus to another form of energy (usually electrical)

Physical phenomena:
Sound
Meter reading
LED indicator
Digital display
Chart recorder
VDU output

3.1 Transfer function

$$\int_{0}^{\infty} \sin \omega t \; e^{-st}$$

3.1.1 Instrumentation

Instrumentation is concerned with producing a measurable output from the signal provided by a transducer. This is usually done through a series of steps or processes, starting with the transducer signal or input, Q_i.

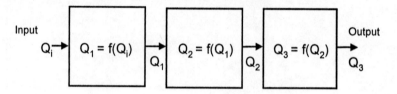

It is desirable to have linear relationships between the inputs and outputs of the various processes so that $Q_3 = K_1 K_2 K_3 Q_i$.

In practice, errors and noise are transmitted at each step along with the signal of interest. The process of instrumentation is to maximise the transmission of signal and to minimise the errors and noise. The process is concerned with:

- the electrical nature of signals and methods of measurement
- signal processing by the application of a **transfer function** to provide amplification and filtering
- the origin and nature of noise in the signal
- signal recovery – filtering, averaging, smoothing, etc.

An important issue in connecting a transducer to a **preamplifier** is to ensure that maximum signal is transferred by a process called **impedance matching.**

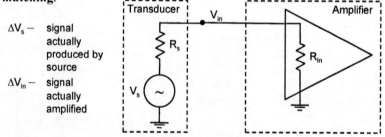

ΔV_s – signal actually produced by source

ΔV_{in} – signal actually amplified

Ideally, $R_{in} \gg R_s$ because otherwise, negligible ΔV_{in} appears at the amplifier. That is, if $R_S \gg R_{in}$, then most of the voltage variations ΔV_S appear across R_s and not at the amplifier input. Amplifiers should have a high input impedance R_{in} compared to R_s, the output resistance of the source.

3.1.2 Transfer function

There is a definite relationship between the input signal S(t) and output
response R(t) of an electronic circuit. The relationship is called the
transfer function.
The transfer function is the ratio of the output voltage over the input
voltage. This is a general definition and includes any phase effects that
might be present in the circuit. For example, for the simple RC low pass
filter shown below, the transfer function is:

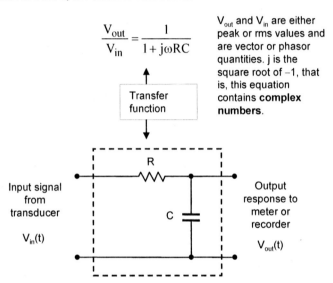

$$\frac{V_{out}}{V_{in}} = \frac{1}{1 + j\omega RC}$$

V_{out} and V_{in} are either
peak or rms values and
are vector or phasor
quantities. j is the
square root of -1, that
is, this equation
contains **complex
numbers**.

Transfer
function

Input signal
from
transducer

$V_{in}(t)$

R

C

Output
response to
meter or
recorder

$V_{out}(t)$

Later, we shall see how the transfer function for a wide variety of circuits
(mainly filter circuits) can be obtained easily using **operator** notation. The
transfer function for a particular filter circuit can be used to modify the
signal being measured so as to eliminate **noise** (i.e. unwanted information).
The mathematical representation of the transfer function allows the effect
of various filter and amplifier circuits to be analysed and designed before
any actual circuit is constructed.

3.1.3 Transforms

Many physical phenomena can be described by **differential equations**, that is, equations which involve derivatives. A short-hand way of writing 'the derivative with respect to' is to use the **differential operator**. For example:

$$\frac{dy}{dx} = Dy \quad \text{and} \quad \frac{d^2y}{dx^2} = D^2y$$

Here, D is a differential operator which, when applied to a function y(x), yields a new function in x. The differential operator may be quite complex, involving derivatives of higher orders. For example:

$$D = a_0 \frac{d^4}{dx^4} + a_1 \frac{d^3}{dx^3} + a_2 \frac{d^2}{dx^2} + a_3 \frac{d}{dx} + a_4 \qquad a_n \text{ are constants}$$

thus

$$Dy = a_0 \frac{d^4y}{dx^4} + a_1 \frac{d^3y}{dx^3} + a_2 \frac{d^2y}{dx^2} + a_3 \frac{dy}{dx} + a_4y$$

After the operator has been applied to the original function, a new function is formed. That is, the original function has been **transformed** into another function.

Original function   Operator   New function

$$y*D = Y$$

The differential operator is very useful for the treatment of many types of differential equations. Another type of transform is the **integral transform operator** T which has the form:

$$T[f(t)] = \int_{-\infty}^{\infty} f(t)K(s,t)\,dt = F(s) \qquad F(s) \text{ is the transform of } f(t)$$

Here, f is a function of t which is transformed by the operator T. K is a function of the variables s and t. The integration produces a new function of s only and is the **integral transform** of the original function f(t). The function K(s,t) can take many forms, and an especially interesting one is that defined by:

$$K(s,t) = 0 \qquad t<0 \qquad \text{The resulting integral transform is called the}$$
$$= e^{-st} \qquad t\geq0 \qquad \textbf{Laplace transform } L[f(t)] \text{ of the function } f(t).$$

3.1.4 Laplace transform

If the function f is a function of t, then the Laplace transform is defined as:

$$L[f(t)] = \int_0^\infty f(t)e^{-st}\,dt$$

The resulting integral, that is, L[f(t)], is a function of s only: L[f(t)] = F(s). F(s) is the Laplace transform of f(t). The symbol 'L' is the **Laplace operator** which acts on f(t) to give the transformed function F(s).

A particularly interesting case is when f(t) is a periodic function, say f(t) =sinωt:

$$L[\sin \omega t] = \int_0^\infty \sin \omega t \; e^{-st}\,dt$$

$$= \frac{\omega}{s^2 + \omega^2} \qquad s > 0$$

The results are shown here without showing the working.

Similarly,

$$L[\cos \omega t] = \frac{s}{s^2 + \omega^2} \qquad s > 0$$

Why are Laplace transforms important to us? Because they allow differential equations to be solved using algebraic expressions involving operators. We'll see how this works in a moment. For now, consider yet another interesting Laplace transform, that of L[1].

$$L[1] = \int_0^\infty e^{-st}\,dt$$

$$= -\frac{1}{s}\left[e^{-st} \right]_0^\infty$$

$$= \frac{1}{s}$$

Well, the question now is, "What is s"? Answer: It depends on the problem being analysed. For sinusoidal signals, it is appropriate to let s = jω.

Why bother with transforms? A particular input signal in the **time domain** may be transformed into another signal, or function, in the **frequency domain**. The transformed signal may then be operated upon by a filter and then transformed back into the time domain for display. The transform of a signal gives information about the composition of the signal.

3.1.5 Operator notation

Consider an **integrator** circuit:

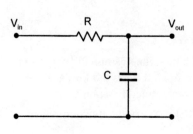

It can be shown that the output voltage is the time integral of the input voltage:

$$V_{out} = \frac{1}{RC} \int V_{in}\, dt$$

when RC is large.

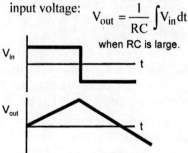

If the input signal is a sine wave, then the output signal is a cosine wave (whose amplitude decreases with increasing frequency of the input signal). It can be shown (see page 218) that the amplitude and phase relationship between V_{in} and V_{out} has the form (in complex number notation):

$$V_{out} = \frac{1}{1 + RCj\omega} V_{in}$$

When RC is large, then:

$$V_{out} \approx \frac{1}{RCj\omega} V_{in}$$

compare

But,

$$V_{out} = \frac{1}{RC} \int V_{in}\, dt$$

Let $\dfrac{d}{dt} = s$

and $\displaystyle\int dt = \frac{1}{s}$

's' is an **operator**. In this case, a "differential" operator since the application of s to a function takes the time derivative of that function.

Thus,

$$V_{out} = \frac{1}{RC} \int V_{in}\, dt$$

$$= \frac{1}{RCs} V_{in}$$

It appears therefore that in this application, $s = j\omega$.

Similarly for a **differentiator**,

$$V_{out} = RC \frac{dV_{in}}{dt}$$

and it can be shown:

$$V_{out} = \frac{1}{1 + \dfrac{1}{RCj\omega}} V_{in}$$

when RC is small.

When RC is small, then, with $s = j\omega$:

$$V_{out} \approx RCj\omega V_{in}$$

$$= RCs V_{in}$$

Is this s the same as that in the Laplace transform? ⟶

3.1.6 Differential operator

Now, if $s = j\omega$, and s also is a **differential operator**, then exactly what is 's' ?

Consider the function:

$$y = e^{-st}$$

$$\frac{dy}{dt} = -se^{-st}$$

$$= -sy$$

That is, 's' *is* a differential operator for this function. Thus:

$$s = \frac{d}{dt}$$

That is, in the **Laplace transform**, where $K(s,t) = e^{-st}$, s can be considered a differential operator and not just an ordinary everyday variable.

For sinusoidal periodic functions, we can let $s = j\omega$ and still have s act like a differential operator. Thus, the s-domain analysis allows us to carry frequency and phase information (for frequency domain analysis) or differential time information (for time domain analysis) in our calculations.

To analyse a circuit, the circuit is transformed into the s-domain, and the necessary algebra performed (which is usually more manageable) and the results transformed back into the frequency or time domain as required.

> This procedure works because in the Laplace transform applied to periodic sinusoidal functions, s is a differential operator *and* is also identified with the product $j\omega$.

On the previous page, we saw how this dual nature of s was consistent with the response of a passive integrator. Let us now apply the technique again to the passive integrator and also the passive differentiator.

3.1.7 Integrator – passive

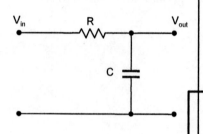

Time domain analysis

$$V_{out} = \frac{Q}{C} = V_C$$

$$V_{in} = V_R + V_C$$

$$\frac{dQ}{dt} = I \therefore Q = \int I dt$$

$$V_{in} = \frac{dQ}{dt}R + \frac{1}{C}Q$$

This is a 1st order 'differential equation' involving differentials with respect to 'time'.

s-domain analysis

$$\frac{d}{dt} = s$$

$$V_{in} = RsQ + \frac{1}{C}Q$$

$$= Q\left(Rs + \frac{1}{C}\right)$$

Now, $V_{out} = \frac{Q}{C}$

thus

$$= \frac{1}{C}\left(\frac{V_{in}}{Rs + \frac{1}{C}}\right)$$

$$\frac{V_{out}}{V_{in}} = \frac{1}{1 + RCs}$$

This equation is a transfer function in s. This is not differential equation, but simple algebraic expression involving the differential operator s. Operator notation allows us to avoid differential equations in the time domain and complex algebra in the frequency domain.

ω-domain analysis

$$V_{in} = IZ$$

$$= I(R - X_C j)$$

$$= I\left(R - \frac{1}{\omega C}j\right)$$

$$V_{out} = I(-X_C j)$$

$$= I\left(-\frac{1}{\omega C}j\right)$$

$$\frac{V_{out}}{V_{in}} = \frac{-\frac{1}{\omega C}j}{R - \frac{1}{\omega C}j}$$

complicated algebra ⬇

$$= \frac{1 - R\omega Cj}{R^2 C^2 \omega^2 + 1}$$

$$= \frac{1}{1 + RC\omega j}$$

let $\quad R\omega C = 1$

then $\quad \dfrac{V_{out}}{V_{in}} = \dfrac{1}{\sqrt{2}}$ $\quad$ 3 dB point

3.1.8 Differentiator – passive

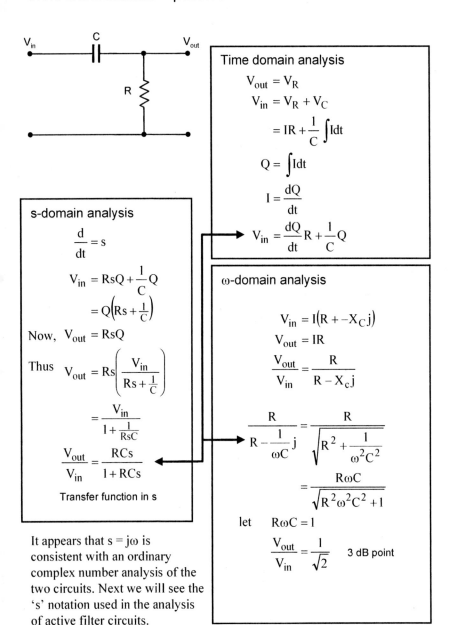

Time domain analysis

$$V_{out} = V_R$$
$$V_{in} = V_R + V_C$$
$$= IR + \frac{1}{C}\int I dt$$
$$Q = \int I dt$$
$$I = \frac{dQ}{dt}$$
$$V_{in} = \frac{dQ}{dt}R + \frac{1}{C}Q$$

s-domain analysis

$$\frac{d}{dt} = s$$

$$V_{in} = RsQ + \frac{1}{C}Q$$
$$= Q\left(Rs + \frac{1}{C}\right)$$

Now, $V_{out} = RsQ$

Thus $V_{out} = Rs\left(\dfrac{V_{in}}{Rs + \frac{1}{C}}\right)$

$$= \frac{V_{in}}{1 + \frac{1}{RsC}}$$

$$\frac{V_{out}}{V_{in}} = \frac{RCs}{1 + RCs}$$

Transfer function in s

ω-domain analysis

$$V_{in} = I(R + -X_C j)$$
$$V_{out} = IR$$
$$\frac{V_{out}}{V_{in}} = \frac{R}{R - X_c j}$$

$$\frac{R}{R - \frac{1}{\omega C}j} = \frac{R}{\sqrt{R^2 + \frac{1}{\omega^2 C^2}}}$$

$$= \frac{R\omega C}{\sqrt{R^2\omega^2 C^2 + 1}}$$

let $\quad R\omega C = 1$

$$\frac{V_{out}}{V_{in}} = \frac{1}{\sqrt{2}} \qquad \text{3 dB point}$$

It appears that s = jω is consistent with an ordinary complex number analysis of the two circuits. Next we will see the 's' notation used in the analysis of active filter circuits.

3.1.9 Transfer impedance

The **transfer impedance** of a network is defined as the ratio of the voltage applied to the input terminals to the current which flows at the output terminals when the output is grounded.

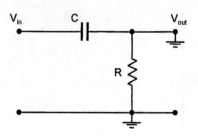

In these simple cases, the transfer impedances are:

$$Z_T = -\frac{1}{\omega C} j$$

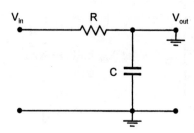

$$Z_T = R$$

3.1.10 Review questions

1. A transducer has an output resistance of 1.2 MΩ. What is the minimum input resistance required for a preamplifier connected to this transducer if at least 95% of the signal emf is to be applied to the preamplifier input?
   (Ans: 23 MΩ)

2. Design a simple RC filter which will attenuate 50 Hz 'hum' by 40 dB. Determine the effect of this filter on the following AC signals: (a) 500 Hz, 0.8 V rms (b) 10 kHz, 1.2 V rms   (Ans: −20 dB, −1 dB)

3. Calculate the centre frequency of the following bandpass filter circuit:

   (Ans: 160 Hz)

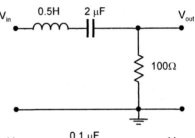

4. The voltage at the input of the following filter is suddenly stepped from 0 to +5 V. Sketch the resulting output voltage as a function of time and calculate the time required for the output to settle to within 0.1 V of its steady state output.

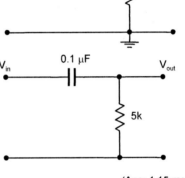

   (Ans: 1.15 msec)

5. Using integration by parts (twice), show that

   $$L[\cos\omega t] = \frac{s}{s^2 + \omega^2}$$

   Hint: $\int u\,dv = uv - \int v\,du$

   $u = e^{-st}$

   $du = -se^{-st}$

   $dv = \cos\omega t\,dt$

   $v = \frac{1}{\omega}\sin\omega t$

6. Using Euler's formula, find L[cosωt]:

   $$e^{j\omega t} = \cos\omega t + j\sin\omega t$$

   (Hint: L[cosωt] is the real part of the expression)

7. If $F(s) = \dfrac{1}{s^2 + 2s + 10}$ find the inverse Laplace transform given that:

   $$L^{-1}(F(s)) = e^{-at}L^{-1}(F(s-a))$$

   and here letting a = 1.

3.1.11 Activities

1. Connect a 741 op-amp to an appropriate power supply and connect the non-inverting input to ground and the inverting input to a variable DC voltage source. Describe what happens at the output when the input voltage is swept from −1 V to +1 V.

2. Sweep the voltage again slowly and determine the input voltage (to within a millivolt) when the output voltage is zero − you may have to modify the circuit slightly to obtain the best estimation of this cross-over voltage.

3. With the non-inverting input still grounded, apply a small AC signal to the inverting input and measure the open-loop gain as a function of frequency. Record your readings in a table, and then plot gain in dB against log of frequency.

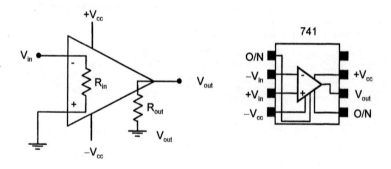

4. Determine the open-loop bandwidth of the op-amp.

5. Construct a simple inverting amplifier with a gain of 100. Check the input offset voltage and connect a nulling circuit to eliminate any offset.

6. Check the voltage at the inverting input and comment on its value.

7. Measure the frequency response of the amplifier. Determine the bandwidth.

8. Measure the input resistance of the 741 IC by altering the circuit to a non-inverting amplifier configuration with a gain of 50.

9. Measure the output resistance by connecting a 100 Ω resistor to the output and to ground and determining the change from open circuit output voltage.

10. Alter the circuit to have a gain of 10 and measure the frequency response and input and output resistances. Compare with previously measured quantities and comment.

Inverting amplifier

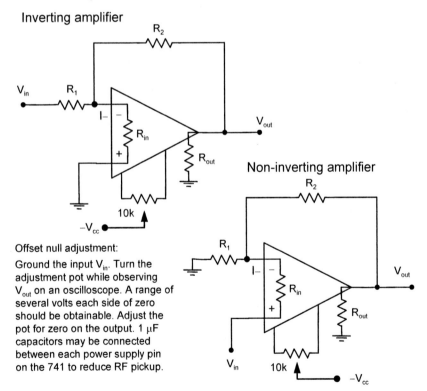

Offset null adjustment:

Ground the input V_{in}. Turn the adjustment pot while observing V_{out} on an oscilloscope. A range of several volts each side of zero should be obtainable. Adjust the pot for zero on the output. 1 μF capacitors may be connected between each power supply pin on the 741 to reduce RF pickup.

Non-inverting amplifier

11. Construct a difference amplifier with a gain of 40.

12. Measure the frequency response and bandwidth of the amplifier.

13. Having measured the difference gain, devise a method to measure the common mode gain of the amplifier and then determine the common mode rejection ratio.

14. Measure the input resistance at each input and comment.

Difference amplifier

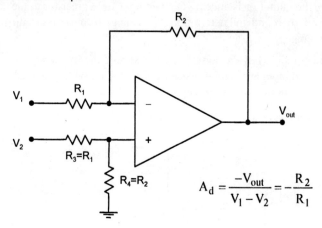

$$A_d = \frac{-V_{out}}{V_1 - V_2} = -\frac{R_2}{R_1}$$

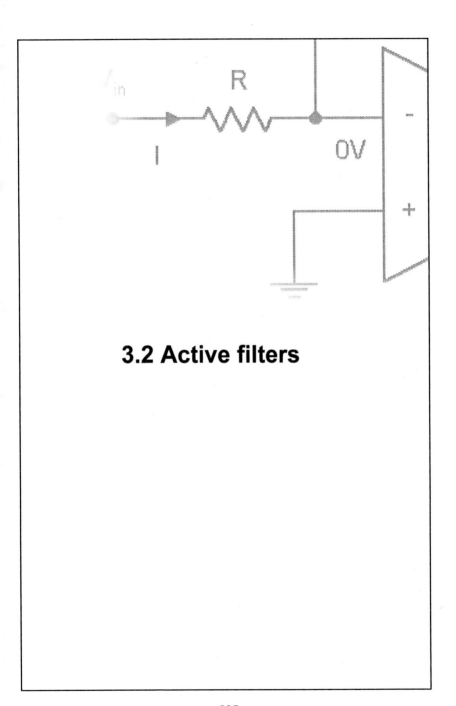

3.2 Active filters

3.2.1 Filters

Passive filters – RLC circuits   1st order filter (1st order differential
equation can describe the filter).

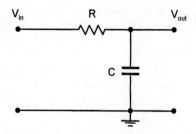

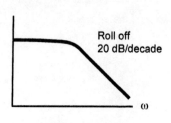

Roll off
20 dB/decade

Two 1st order filters cascaded together produce a 2nd order filter:

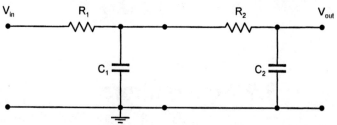

Input Z of 2nd stage = output Z of 1st
stage. To minimise loading, $R_2 \gg R_1$
and $C_2 \ll C_1$.

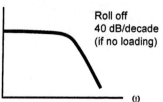

Roll off
40 dB/decade
(if no loading)

Active filters – op-amp circuits

Advantages:
- can incorporate gain
- loading is not such a problem
- cascading of filters for 2nd
  order
- don't have to use inductors
  (which are expensive, require
  large currents, generate back
  emfs)
- tuning of filters can be done by
  adjusting resistors

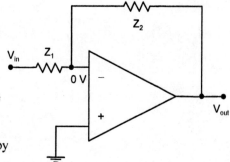

3.2.2 T -network filters

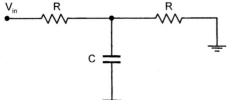

The **transfer impedance** of a network is defined as the ratio of the voltage applied to the input terminals to the current which flows at the output terminals when the output is grounded.

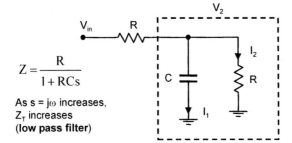

$$Z = \frac{R}{1 + RCs}$$

As $s = j\omega$ increases,
Z_T increases
(low pass filter)

Now,

$V_2 = IZ$

$= I \dfrac{R}{1 + RCs}$ with RHS grounded

$= I_2 R$

$I = I_2 R \dfrac{1 + RCs}{R}$

$V_{in} = IR + IZ$

$= I(R + Z)$

$= I_2 R \dfrac{1 + RCs}{R} \left(R + \dfrac{R}{1 + RCs} \right)$

$= I_2 R (1 + RCs) \left(1 + \dfrac{1}{1 + RCs} \right)$

$\dfrac{V_{in}}{I_2} = R (1 + RCs) \left(1 + \dfrac{1}{1 + RCs} \right) = Z_T$

$Z_T = R (2 + RCs)$

Transfer impedance

A similar analysis for a **high pass** network yields

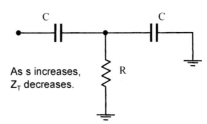

As s increases,
Z_T decreases.

$$Z_T = \frac{V_{in}}{I_2} = \frac{1}{Cs} \left(\frac{1 + 2RCs}{RCs} \right)$$

Texts on this subject often provide tables of transfer impedances for standard network arrangements which can then be easily combined (in the s-domain) for a particular application.

3.2.3 Twin-T filter

Tuned **rejection** filter

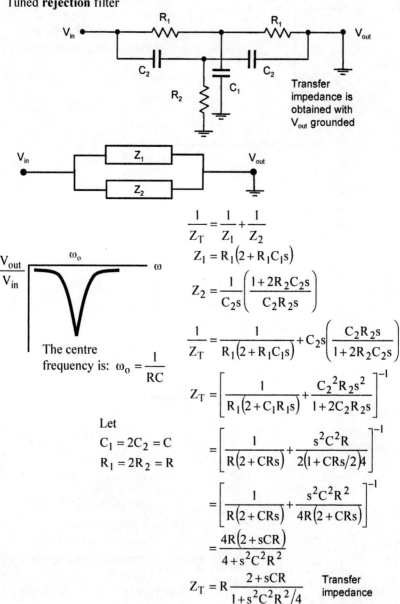

$$\frac{1}{Z_T} = \frac{1}{Z_1} + \frac{1}{Z_2}$$

$$Z_1 = R_1(2 + R_1 C_1 s)$$

$$Z_2 = \frac{1}{C_2 s}\left(\frac{1 + 2R_2 C_2 s}{C_2 R_2 s}\right)$$

$$\frac{1}{Z_T} = \frac{1}{R_1(2 + R_1 C_1 s)} + C_2 s\left(\frac{C_2 R_2 s}{1 + 2R_2 C_2 s}\right)$$

$$Z_T = \left[\frac{1}{R_1(2 + C_1 R_1 s)} + \frac{C_2{}^2 R_2 s^2}{1 + 2C_2 R_2 s}\right]^{-1}$$

The centre frequency is: $\omega_o = \dfrac{1}{RC}$

Let
$$C_1 = 2C_2 = C$$
$$R_1 = 2R_2 = R$$

$$= \left[\frac{1}{R(2 + CRs)} + \frac{s^2 C^2 R}{2(1 + CRs/2)4}\right]^{-1}$$

$$= \left[\frac{1}{R(2 + CRs)} + \frac{s^2 C^2 R^2}{4R(2 + CRs)}\right]^{-1}$$

$$= \frac{4R(2 + sCR)}{4 + s^2 C^2 R^2}$$

$$Z_T = R\frac{2 + sCR}{1 + s^2 C^2 R^2/4} \qquad \text{Transfer impedance}$$

3.2.4 Active integrator/differentiator

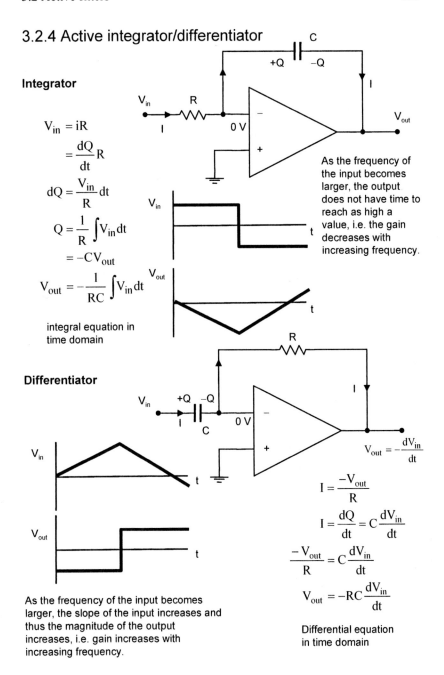

Integrator

$$V_{in} = iR$$

$$= \frac{dQ}{dt} R$$

$$dQ = \frac{V_{in}}{R} dt$$

$$Q = \frac{1}{R} \int V_{in} dt$$

$$= -CV_{out}$$

$$V_{out} = -\frac{1}{RC} \int V_{in} dt$$

integral equation in
time domain

As the frequency of
the input becomes
larger, the output
does not have time to
reach as high a
value, i.e. the gain
decreases with
increasing frequency.

Differentiator

$$V_{out} = -\frac{dV_{in}}{dt}$$

$$I = \frac{-V_{out}}{R}$$

$$I = \frac{dQ}{dt} = C \frac{dV_{in}}{dt}$$

$$\frac{-V_{out}}{R} = C \frac{dV_{in}}{dt}$$

$$V_{out} = -RC \frac{dV_{in}}{dt}$$

Differential equation
in time domain

As the frequency of the input becomes
larger, the slope of the input increases and
thus the magnitude of the output
increases, i.e. gain increases with
increasing frequency.

3.2.5 Integrator transfer function

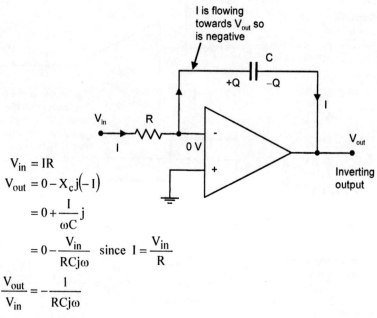

$$V_{in} = IR$$
$$V_{out} = 0 - X_c j(-I)$$
$$= 0 + \frac{I}{\omega C} j$$
$$= 0 - \frac{V_{in}}{RCj\omega} \quad \text{since } I = \frac{V_{in}}{R}$$
$$\frac{V_{out}}{V_{in}} = -\frac{1}{RCj\omega}$$

Transfer function in
ω-domain

Now, $s = j\omega$

$$\frac{V_{out}}{V_{in}} = -\frac{1}{RCj\omega}$$
$$= -\frac{1}{RCs} \quad \text{Transfer function in s-domain}$$
$$V_{out} = -\frac{1}{RC}\frac{1}{s}V_{in}$$

but $\quad \int dt = \frac{1}{s}$

thus $\quad V_{out} = -\frac{1}{RC}\int V_{in}\, dt$

General transfer function

Now $\quad Z_f = X_c = \frac{1}{Cj\omega}$

and $\quad Z_i = R$

thus $\quad \dfrac{V_{out}}{V_{in}} = \left[-\dfrac{Z_f}{Z_i} \right]$

This is a general transfer function
that holds for a general circuit with
feedback elements.

3.2.6 Low pass filter – active

An active integrator can be modified to act like a low pass filter.

Without R_2, at low frequency, the **gain** (V_{out}/V_{in}) becomes very large and at DC, approaches the open-loop gain. Need a low frequency cutoff to eliminate drift. This is the function of R_2.

$$V_{out} = \left[-\frac{1}{RCj\omega} \right] V_{in}$$

$$Z_i = R; \quad Z_f = \left[\frac{1}{R_2} + sC \right]^{-1}$$

$$= \frac{R_2}{1 + R_2 Cs} \qquad \text{Let } s = j\omega$$

since $V_{out} = \left[-\dfrac{Z_f}{Z_i} \right] V_{in}$

then $\dfrac{V_{out}}{V_{in}} = -\dfrac{R_2}{R_1} \dfrac{1}{1 + R_2 Cs}$

If s is small, then:

$$V_{out} = -\frac{R_2}{R_1} V_{in}$$

If s is large, then

$$V_{out} = -\frac{1}{R_1 Cs} V_{in}$$

Transfer function in s-domain

Integrator

$\dfrac{V_{out}}{V_{in}}$

$\dfrac{R_2}{R_1}$

$|R_2Cs| = 1$

Frequency domain analysis

$$V_{out} = -\frac{R_2}{R_1} \frac{1}{1 + R_2 sC} V_{in}$$

$$= -\frac{R_2}{R_1} \frac{1}{1 + R_2 Cj\omega} V_{in}$$

$$= -\frac{R_2}{R_1} \frac{1 - R_2 Cj\omega}{1 + [R_2 C\omega]^2} V_{in}$$

$$\left| V_{out} \right| = \frac{R_2}{R_1} \frac{\left(1 + [R_2 C\omega]^2 \right)^{\frac{1}{2}}}{1 + [R_2 C\omega]^2} \left| V_{in} \right|$$

$$\left| \frac{V_{out}}{V_{in}} \right| = \frac{R_2}{R_1} \frac{1}{\left(1 + [R_2 \omega C]^2 \right)^{\frac{1}{2}}}$$

Transfer function in ω-domain

at $R_2 \omega C = 1$

$$\left| \frac{V_{out}}{V_{in}} \right| = \frac{R_2}{R_1} \frac{1}{\sqrt{2}}$$

Keep in mind that $s = j\omega$ hence the use of | |. Need to square $R_2 sC$ and then take the square root to find the magnitude.

Note, compared to the passive integrator, this circuit contains an element of "gain" equal to the ratio R_2/R_1.

3.2.7 2nd order active filter

Low pass

$$I_1 = I_2 + I_3 + I_4$$

$$I_1 = \frac{V_{in} - V_N}{R}$$

$$I_2 = \frac{V_N}{\frac{1}{C_1 s}} = V_N C_1 s$$

$$I_3 = \frac{-V_{out}}{\frac{1}{C_2 s}} = -V_{out} C_2 s$$

$$I_4 = \frac{V_N - V_{out}}{R}$$

Note: For a negative V_{out}, then I_3 flows in the direction indicated and is thus positive.

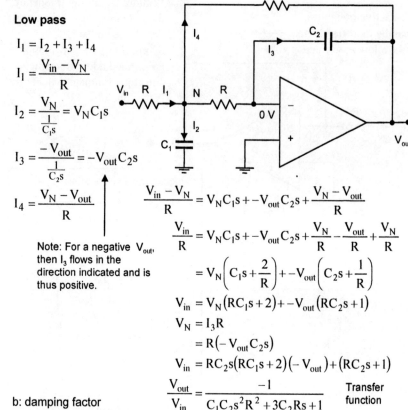

$$\frac{V_{in} - V_N}{R} = V_N C_1 s + -V_{out} C_2 s + \frac{V_N - V_{out}}{R}$$

$$\frac{V_{in}}{R} = V_N C_1 s + -V_{out} C_2 s + \frac{V_N}{R} - \frac{V_{out}}{R} + \frac{V_N}{R}$$

$$= V_N \left(C_1 s + \frac{2}{R} \right) + -V_{out} \left(C_2 s + \frac{1}{R} \right)$$

$$V_{in} = V_N (RC_1 s + 2) + -V_{out} (RC_2 s + 1)$$

$$V_N = I_3 R$$

$$= R(-V_{out} C_2 s)$$

$$V_{in} = RC_2 s(RC_1 s + 2)(-V_{out}) + (RC_2 s + 1)$$

$$\frac{V_{out}}{V_{in}} = \frac{-1}{C_1 C_2 s^2 R^2 + 3C_2 Rs + 1} \qquad \text{Transfer function}$$

b: damping factor

b = 2 critically damped

b > 2 overdamped

b < 2 underdamped

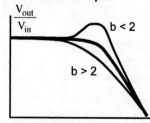

Letting $\quad C_1 = \dfrac{3}{b} C$

and $\quad C_2 = \dfrac{b}{3} C$

$$\frac{V_{out}}{V_{in}} = \frac{-1}{C_1 C_2 s^2 R^2 + 3C_2 Rs + 1}$$

$$= \frac{-1}{R^2 C^2 s^2 + RCsb + 1}$$

The analysis of this circuit in the frequency or time domain would be very cumbersome. Here we have arrived at a transfer function in the s-domain with very little effort.

3.2.8 Double integrator

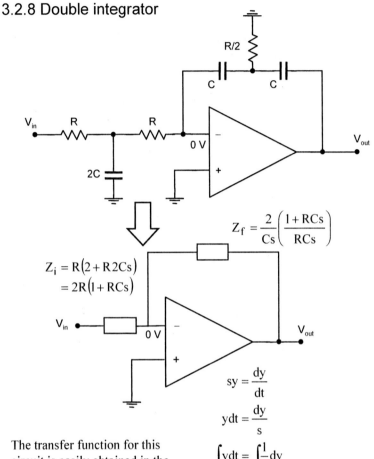

$$Z_f = \frac{2}{Cs}\left(\frac{1+RCs}{RCs}\right)$$

$$Z_i = R(2+R2Cs)$$
$$= 2R(1+RCs)$$

$$sy = \frac{dy}{dt}$$

$$ydt = \frac{dy}{s}$$

The transfer function for this circuit is easily obtained in the s-domain using transfer impedance results for the T–networks shown previously. The final transfer function involves a s^{-2} term. Now, s is a differential operator so that s^{-1} is an integral operator. Here, we have an s^{-2} which means that the circuit takes the integral twice – a **double integrator**.

$$\int ydt = \int \frac{1}{s}dy$$
$$= \frac{1}{s}y$$

$$\frac{V_{out}}{V_{in}} = \left[-\frac{Z_f}{Z_i}\right]$$

$$= -\frac{2}{Cs}\left(\frac{1+RCs}{RCs}\right)\frac{1}{2R(1+RCs)}$$

$$= -\frac{1}{R^2C^2s^2}$$

3.2.9 Bandpass filter – narrow

An active filter with a twin–T
network as the feedback element

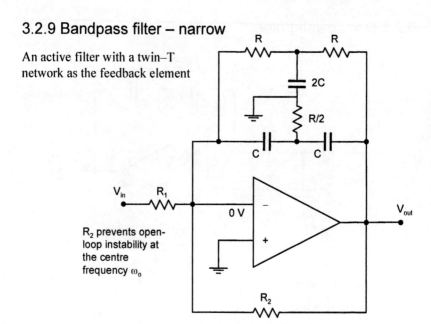

R_2 prevents open-
loop instability at
the centre
frequency ω_0

When twin-T is used as the feedback element, Z_T is very high at the centre
frequency ω_0 and thus the gain of the active op-amp circuit is a maximum.
At other frequencies, Z_T is small and thus the gain of the overall circuit is a
minimum. This leads to a **bandpass** filter with the following
characteristics:

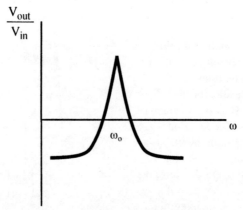

If the twin-T filter moved to the input, we would obtain a **band reject** or
notch filter.

3.2.10 Differentiator transfer function

$$V_{in} = IZ_i$$

$$= \frac{-I}{\omega C} j$$

$$= \frac{I}{Cj\omega}$$

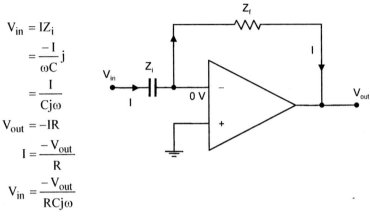

$$V_{out} = -IR$$

$$I = \frac{-V_{out}}{R}$$

$$V_{in} = \frac{-V_{out}}{RCj\omega}$$

$$\frac{V_{out}}{V_{in}} = -RCj\omega \qquad \text{Transfer function in } \omega\text{-domain}$$

Now, $s = j\omega$

$$\frac{V_{out}}{V_{in}} = -RCj\omega$$

$$= -RCs \qquad \text{Transfer function in s-domain}$$

but $\quad s = \dfrac{d}{dt}$

thus $\quad V_{out} = -RC\dfrac{dV_{in}}{dt}$

> Now $\quad Z_f = R$
>
> and $\quad Z_i = \dfrac{1}{j\omega C}$
>
> thus $\quad \dfrac{Z_f}{Z_i} = RCj\omega$
>
> $\quad \dfrac{V_{out}}{V_{in}} = -\dfrac{Z_f}{Z_i} \qquad$ General transfer function

3.2.11 High pass filter – active

With an active differentiator, as the frequency increases, the output voltage
increases without limit in the ideal case (actually limited by the V+ and V–
supply). This is undesirable since high frequency noise will be greatly
amplified. The solution is to build in a **cutoff** frequency. The circuit then
acts like a high pass filter.

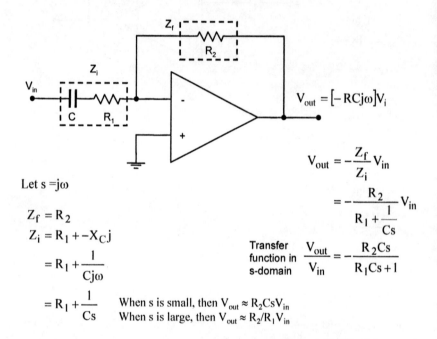

$$V_{out} = [-RCj\omega]V_i$$

$$V_{out} = -\frac{Z_f}{Z_i}V_{in}$$

Let $s = j\omega$

$$Z_f = R_2$$
$$Z_i = R_1 + -X_C j$$
$$= R_1 + \frac{1}{Cj\omega}$$

$$= -\frac{R_2}{R_1 + \dfrac{1}{Cs}}V_{in}$$

Transfer function in s-domain

$$\frac{V_{out}}{V_{in}} = -\frac{R_2 Cs}{R_1 Cs + 1}$$

$$= R_1 + \frac{1}{Cs}$$

When s is small, then $V_{out} \approx R_2 Cs V_{in}$
When s is large, then $V_{out} \approx R_2/R_1 V_{in}$

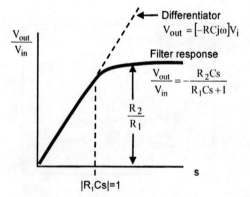

Differentiator
$$V_{out} = [-RCj\omega]V_i$$

Filter response
$$\frac{V_{out}}{V_{in}} = -\frac{R_2 Cs}{R_1 Cs + 1}$$

If we multiply and
divide the numerator by
R_1, then we have:

$$\frac{V_{out}}{V_{in}} = -\frac{R_2}{R_1}\frac{R_1 sC}{R_1 sC + 1}$$

Letting $|R_1 sC| = 1$ gives us
the **3 dB point**. Note the
element of gain R_2/R_1
compared to the passive
differentiator.

3.2.12 High pass filter – ω-domain

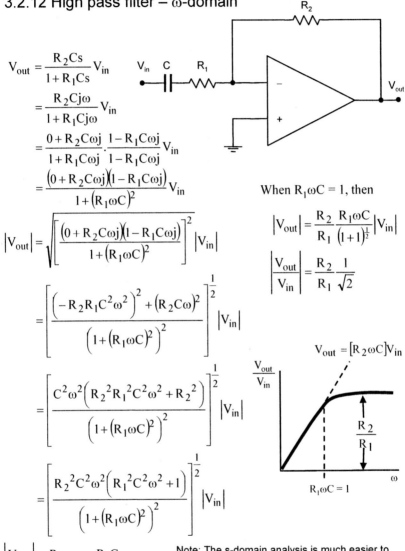

$$V_{out} = \frac{R_2Cs}{1+R_1Cs}V_{in}$$

$$= \frac{R_2Cj\omega}{1+R_1Cj\omega}V_{in}$$

$$= \frac{0+R_2C\omega j}{1+R_1C\omega j}\cdot\frac{1-R_1C\omega j}{1-R_1C\omega j}V_{in}$$

$$= \frac{(0+R_2C\omega j)(1-R_1C\omega j)}{1+(R_1\omega C)^2}V_{in}$$

$$|V_{out}| = \sqrt{\left[\frac{(0+R_2C\omega j)(1-R_1C\omega j)}{1+(R_1\omega C)^2}\right]^2}|V_{in}|$$

$$= \left[\frac{\left(-R_2R_1C^2\omega^2\right)^2+(R_2C\omega)^2}{\left(1+(R_1\omega C)^2\right)^2}\right]^{\frac{1}{2}}|V_{in}|$$

$$= \left[\frac{C^2\omega^2\left(R_2{}^2R_1{}^2C^2\omega^2+R_2{}^2\right)}{\left(1+(R_1\omega C)^2\right)^2}\right]^{\frac{1}{2}}|V_{in}|$$

$$= \left[\frac{R_2{}^2C^2\omega^2\left(R_1{}^2C^2\omega^2+1\right)}{\left(1+(R_1\omega C)^2\right)^2}\right]^{\frac{1}{2}}|V_{in}|$$

When $R_1\omega C = 1$, then

$$|V_{out}| = \frac{R_2}{R_1}\frac{R_1\omega C}{(1+1)^{\frac{1}{2}}}|V_{in}|$$

$$\left|\frac{V_{out}}{V_{in}}\right| = \frac{R_2}{R_1}\frac{1}{\sqrt{2}}$$

$$V_{out} = [R_2\omega C]V_{in}$$

$$\left|\frac{V_{out}}{V_{in}}\right| = \frac{R_2}{R_1}\frac{R_1C\omega}{\sqrt{\left(1+(R_1\omega C)^2\right)}}$$

Transfer function in ω-domain

Note: The s-domain analysis is much easier to handle. It allows the general characteristics of a circuit to be readily analysed. The precise shape of the frequency response of a circuit needs to be obtained, however, from the ω-domain analysis.

3.2.13 Bandpass filter – wide

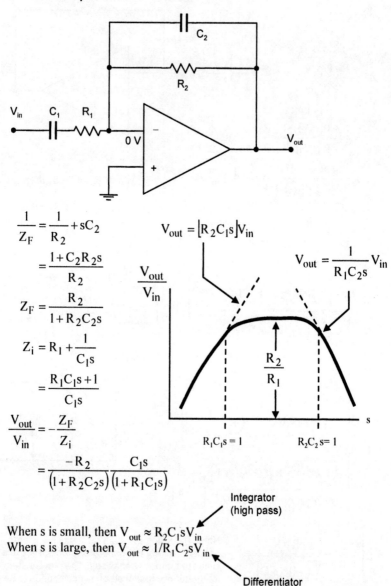

$$\frac{1}{Z_F} = \frac{1}{R_2} + sC_2$$

$$= \frac{1 + C_2 R_2 s}{R_2}$$

$$Z_F = \frac{R_2}{1 + R_2 C_2 s}$$

$$Z_i = R_1 + \frac{1}{C_1 s}$$

$$= \frac{R_1 C_1 s + 1}{C_1 s}$$

$$\frac{V_{out}}{V_{in}} = -\frac{Z_F}{Z_i}$$

$$= \frac{-R_2}{(1 + R_2 C_2 s)} \frac{C_1 s}{(1 + R_1 C_1 s)}$$

$$V_{out} = [R_2 C_1 s] V_{in}$$

$$V_{out} = \frac{1}{R_1 C_2 s} V_{in}$$

Integrator
(high pass)

When s is small, then $V_{out} \approx R_2 C_1 s V_{in}$
When s is large, then $V_{out} \approx 1/R_1 C_2 s V_{in}$

Differentiator
(low pass)

3.2.14 Voltage gain and dB

The voltage gain of a circuit is obtained from the transfer function:

$$A_v = \frac{V_{out}}{V_{in}} = -\frac{Z_f}{Z_i}$$

and may be expressed as a ratio (e.g. $A_v = 100$). However, the gain of a circuit may cover several orders of magnitude depending on the frequency of the input signal. To facilitate this range of possible values of gain, it is often more convenient to express gain on a logarithmic scale. The scale chosen is the "**decibel**" scale (really a power gain).

$$db = 10\log\left(\frac{V_o}{V_i}\right)^2$$

The square factor is applied because the decibel scale represents the "power" output of a circuit and:

$$= 20\log\frac{V_o}{V_i}$$

$$P = \frac{V^2}{R}$$

3.2.15 Review questions

1. Consider the filter circuit:

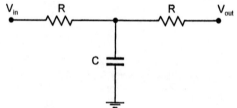

(a) Determine the transfer function of the circuit. (Hint: Write the differential equation relating the input and output voltages with respect to time.)

(b) Under what circumstances may the circuit be used as an analog integrator?

2. Determine the s-domain transfer impedance for the T-type network shown below.

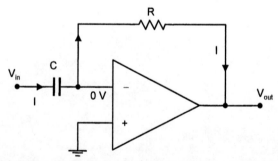

3. (a) Show how the circuit below acts as an analog differentiator.

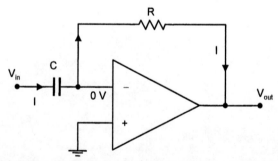

(b) The differentiator shown above is susceptible to high frequency noise. Explain why this occurs.

(c) Modify the circuit to reduce high frequency noise and determine the transfer function of this modified circuit.

(d) Sketch the transfer functions of the original and modified circuits and discuss their features.

4. Consider the modified integrator circuit:

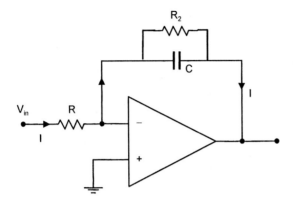

(a) What is the function of the feedback resistor R_2?

(b) Derive an expression for the transfer function in the s-domain.

(c) Select values of resistors and capacitors to give integration of signals above 50 Hz.

3.2.16 Activities

1. Construct the 1st order and 2nd order low pass filters as shown. Choose component values to give cutoff frequencies of a few kilohertz.

2. Using a sinusoidal input signal, measure the frequency response of the filter circuits and plot the transfer function of each filter. Compare the 3 dB points and roll-off for each filter.

3. Using a square wave input, examine the step response of each filter. Compare the responses of the two filters.

4. With the 2nd order filter, alter the value of the parameter b and examine its effect on the step response of the circuit.

1st order

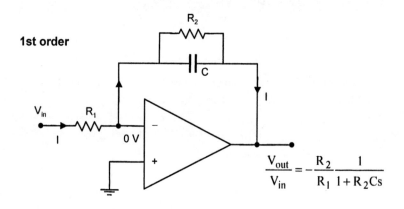

$$\frac{V_{out}}{V_{in}} = -\frac{R_2}{R_1}\frac{1}{1+R_2Cs}$$

2nd order

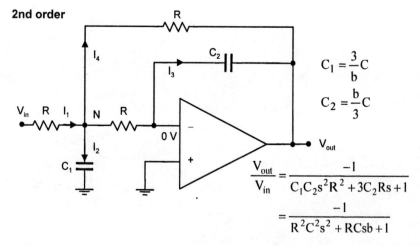

$$C_1 = \frac{3}{b}C$$

$$C_2 = \frac{b}{3}C$$

$$\frac{V_{out}}{V_{in}} = \frac{-1}{C_1C_2s^2R^2 + 3C_2Rs + 1}$$

$$= \frac{-1}{R^2C^2s^2 + RCsb + 1}$$

5. Design a twin-T bandpass filter to have a centre frequency of $\omega_o =$ 1000Hz.

6. Construct the twin-T network and measure its transfer characteristics. Plot on an appropriate graph.

7. Using this twin-T network, construct a bandpass filter. Adjust R_2 for stability if necessary.

8. Plot the transfer function of this filter and comment on the significant features.

9. Determine the theoretical transfer function of this circuit and compare with that measured.

10. Examine the step response of this filter and comment.

Twin-T network

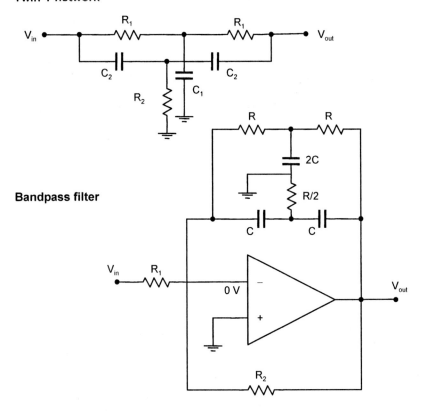

Bandpass filter

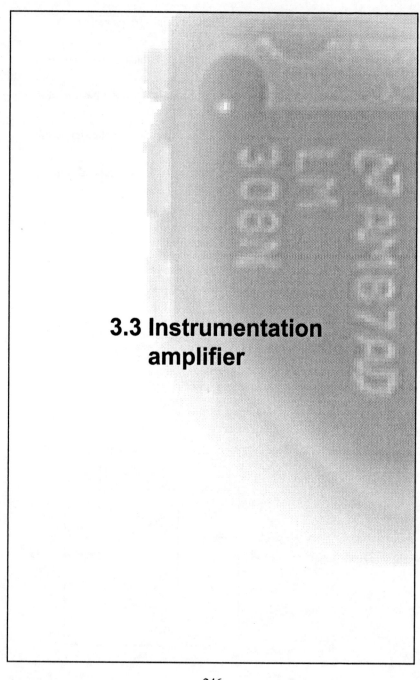

3.3 Instrumentation amplifier

3.3.1 Difference amplifier

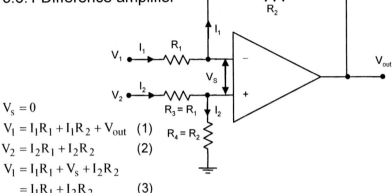

$V_s = 0$

$V_1 = I_1 R_1 + I_1 R_2 + V_{out}$ (1)

$V_2 = I_2 R_1 + I_2 R_2$ (2)

$V_1 = I_1 R_1 + V_s + I_2 R_2$

$\quad = I_1 R_1 + I_2 R_2$ (3)

Note: The negative input to the op-amp is not a virtual earth (0 V) in this circuit. The internal input resistance of the op-amp is MΩ but because the input bias currents are nA, $V_s = 0$ and so the voltage at negative input is equal to $I_2 R_2$.

from (1)

$$V_1 = I_1 R_1 + I_1 R_2 + V_{out}$$
$$V_1 - V_{out} = I_1 R_1 + I_1 R_2$$
$$I_1 = \frac{V_1 - V_{out}}{R_1 + R_2}$$

substituting into (3)

$$V_1 = I_1 R_1 + I_2 R_2$$
$$= \frac{(V_1 - V_{out})}{R_1 + R_2} R_1 + \frac{V_2}{R_1 + R_2} R_2$$
$$V_1 (R_1 + R_2) = (V_1 - V_{out}) R_1 + V_2 R_2$$
$$V_1 R_1 + V_2 R_2 = V_1 R_1 - V_{out} R_1 + V_2 R_2$$
$$V_1 R_2 = V_2 R_2 - V_{out} R_1$$
$$(V_1 - V_2) R_2 = -V_{out} R_1$$
$$-V_{out} = \frac{R_2}{R_1} (V_1 - V_2)$$

$$A_d = \frac{-V_{out}}{V_1 - V_2} = -\frac{R_2}{R_1}$$

difference gain

from (2)

$$V_2 = I_2 R_1 + I_2 R_2$$
$$= I_2 (R_1 + R_2)$$
$$I_2 = \frac{V_2}{R_1 + R_2}$$

Now, the input resistance of the negative input is unbalanced with respect to the positive input. If V_1 is grounded, then R_{in} of V_2 is $R_1 + R_2$. If V_2 is grounded, then R_{in} of V_1 is R_1 (since when V_2 is grounded, the negative input *is* a virtual earth). This can cause problems with uneven loading of the sources. To overcome this, we can design an input stage using **voltage followers**.

3.3.2 CMRR

The **common mode rejection ratio** (CMRR) is the ratio of the differential gain to the common mode gain. The common mode gain is that obtained when $V_1 = V_2$

$$CMRR = \frac{A_d}{A_{cm}} = 20\log_{10}\frac{A_d}{A_{cm}}$$

The more general expression for difference gain is:

$$-V_{out} = \frac{R_4}{R_1}\left(\frac{R_1 + R_2}{R_3 + R_4}\right)V_2 - \frac{R_2}{R_1}V_1$$

With a common mode signal, $V_1 = V_2$, thus:

$$\frac{-V_{out}}{V_{in}} = \frac{R_4}{R_1}\left(\frac{R_1 + R_2}{R_3 + R_4}\right) - \frac{R_2}{R_1}$$

$$= A_{cm}$$

Small variations in resistor values in a circuit can lead to some common mode gain.

Now consider the following circuit where the source voltages and output resistances are included:

$$V_{S1} = I_1(R_S + R_1)$$
$$V_1 = I_1 R_1$$
$$V_1 = \frac{R_1 V_{S1}}{R_S + R_1}$$
$$V_{S2} = I_2(R_S + R_3 + R_4)$$
$$V_2 = I_2(R_3 + R_4)$$
$$V_2 = \frac{(R_3 + R_4)V_{S2}}{R_S + R_3 + R_4}$$
$$= \frac{(R_1 + R_2)V_{S2}}{R_S + R_1 + R_2}$$
for matched resistors

Now, even if $V_{S1} = V_{S2}$ and resistors are matched, $V_1 \neq V_2$ and thus some common mode gain is the result. The difference in V_1 and V_2 gets smaller as R_S is reduced. At $R_S = 0$, $V_1 = V_2 = V_S$ and no common mode gain. For the highest common mode rejection ratio, the amplifier should be driven by low impedance sources – such as a voltage follower.

3.3.3 Difference amplifier with voltage follower inputs

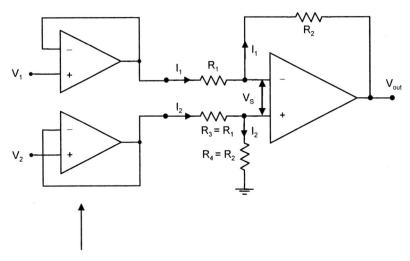

Unity gain ($\beta = 1$) voltage followers: high input impedance, low output impedance.

$$R_{in\,new} = R_{in\,old}\left(1 + |\beta|A_o\right)$$

$$\approx R_{in\,old}\,10^5$$

$\beta = 1$

$A_o = 10^5$

Signal sources see only high impedances, therefore maximum transfer of V_s and no uneven loading of the sources.

The amplifier itself is driven by low impedance sources (R_{out} of an op-amp is very small: 75 Ω). CMRR is improved. The effect on CMRR of source impedance is much greater than resistance mismatches.

3.3.4 Difference amplifier with cross-coupled inputs

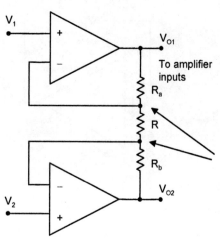

Note: This input stage **is not a difference amplifier**. The difference in the output voltages = the gain times the difference in the input voltages. Common mode signals are passed through without being amplified. A proper difference amplifier rejects the common mode signal altogether.

To amplifier inputs

Feedback resistors R_a and R_b tend to keep the negative and positive inputs to the op-amp at equal potential hence voltages at R are V_1 and V_2

Now, $V_1 - V_2 = IR$

$$V_{o1} - V_1 = IR_a$$

$$V_2 - V_{o2} = IR_b$$

thus $$I = \frac{V_1 - V_2}{R}$$

$$V_{o1} - V_1 = \frac{V_1 - V_2}{R} R_a$$

$$V_2 - V_{o2} = \frac{V_1 - V_2}{R} R_b$$

$$V_{o1} - V_1 + V_2 - V_{o2} = \frac{V_1 - V_2}{R}\left(R_a + R_b\right)$$

$$V_{o1} - V_{o2} = \frac{V_1 - V_2}{R}\left(R_a + R_b\right) + \left(V_1 - V_2\right)$$

This is the **gain** of the input stage. ⟶ $$A_i = \frac{\left(R_a + R_b + R\right)}{R}$$

The gain of the input stage can thus be altered by adjusting just one resistor R.

Gain increases as R decreases. If R is made very large, the gain approaches 1

3.3.5 CMRR cross-coupled inputs

If the input signal consists of a
common mode component,
(e.g. $V_1 = 5$ V, $V_2 = 3$ V means that
$V_{cm} = 3$V and $V_1 - V_2 = 2$V) then:

$$V_1 = V_{cm} - \frac{V_1 - V_2}{2}$$

$$V_2 = V_{cm} + \frac{V_1 - V_2}{2}$$

$$\frac{V_1 - V_2}{2} = V_2 - V_{cm}$$

$$V_{cm} = V_1 + V_2 - V_{cm}$$

$$V_{cm} = \frac{V_1 + V_2}{2}$$

$$V_{cm} = \frac{V_1 + V_2}{2}$$

Since common mode signals
are not amplified, then:

$$Vo_{cm} = \frac{V_{o1} + V_{o2}}{2}$$

thus $\quad A_{cm} = \frac{(V_{o1} + V_{o2})/2}{(V_1 + V_2)/2}$

$$= \frac{V_{o1} + V_{o2}}{V_1 + V_2}$$

but $\quad V_1 - V_2 = IR$

$$V_{o1} = IR_a + V_1$$

$$= \frac{V_1 - V_2}{R} R_a + V_1$$

and $\quad V_{o2} = V_2 - IR_b$

$$= V_2 - \frac{V_1 + V_2}{R} R_b$$

if $\quad R_a = R_b$

then $V_{o1} + V_{o2} = \frac{V_1 - V_2}{R} R_a + V_1 + V_2 - \frac{V_1 - V_2}{R} R_a$

$$= V_1 + V_2$$

therefore $A_{cm} = 1$

Now, $CMRR = \dfrac{A_i}{A_{cm}}$

but $\quad A_{cm} = 1$

thus $\quad CMRR = A_i$

But, A_i is the gain of the input
stage which is adjustable via R.
This means that the CMRR is
adjustable. For highest CMRR
we thus require a high value of
A_i (and hence a low value of R).

3.3.6 Instrumentation amplifier

An **instrumentation amplifier** is characterised by a high gain and high CMRR.

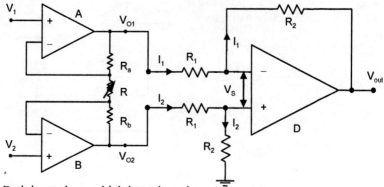

- Both inputs have a high input impedance.
- The gain of the amplifier can be easily adjusted via R.
- The resistors R_1 at the input to the final differential amplifier are trimmed to eliminate amplification of any common mode signal.

The gain of the input stage is:

$$V_{o1} - V_{o2} = \frac{V_1 - V_2}{R}\left(R_a + R_b + R\right)$$

The gain of the amplifier stage is:

$$A_d = \frac{R_2}{R_1}$$

It is usual to have the required gain of the overall circuit obtained from the input stage and the R_2/R_1 term drops out. The difference amplifier D is designed for a gain of 1 and its purpose is to reject any common mode signal.

Thus the total gain is the product of the two:

$$A_v = -\frac{V_1 - V_2}{R}\left(R_a + R_b + R\right)\frac{R_2}{R_1}$$

$$= -\frac{V_1 - V_2}{R}\left(R_a + R_a + R\right)\frac{R_2}{R_1} \quad \text{letting } R_a = R_b$$

$$A_v = -\left(\frac{2R_a}{R} + 1\right)\frac{R_2}{R_1}\left(V_1 - V_2\right)$$

3.3.7 Log amplifier

A non-linear resistor is connected into the feedback circuit. In practice, this can be a diode, but a transistor connected as a diode is used since the forward biased transfer function is more accurately exponential. The exponential nature of the forward biased **diode** leads to a logarithmic decrease in gain of the circuit as the input signal is increased.

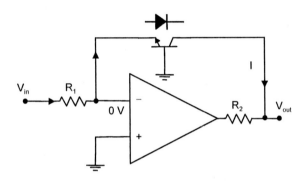

The **feedback transistor** has its collector at 0 V (virtual earth) and the base is also at ground potential (0 V). With the collector and base effectively shorted together, the device acts like a diode across the base–emitter pn junction.

$$V_{in} = IR_1$$

$$I = I_0 e^{eV_{out}/kT}$$

Note: $e/kT \approx 40$ at T = 300K

$$\frac{V_{in}}{R_1 I_0} = e^{eV_{out}/kT}$$

$e = 1.6 \times 10^{-19}$C
$k = 1.38 \times 10^{-23}$ J/K

$$\ln\frac{V_{in}}{R_1 I_0} = \frac{eV_{out}}{kT}$$

Transfer function $$V_{out} = \frac{kT}{e}\ln\frac{V_{in}}{R_1 I_0}$$

$$V_{out} \approx 0.026 \ln V_{in} - 0.026 \ln(R_1 I_0)$$

This amplifier has a high gain for small signals (low V_{in}) and a (logarithmic) progressively lower gain for increasing signals.

The forward bias transfer function of the diode is given by the diode equation:

$$I \approx I_0 e^{eV/kT}$$

where I_0 is the reverse bias leakage current.

Note: This approximation holds for forward bias where $I \gg I_0$. Thus, the transfer function shown here requires V_{in} to be positive so that the pn junction is always well into forward bias.

3.3.8 Op-amp frequency response

Thus far, it has been assumed that the op-amp has an infinite bandwidth and that the frequency response of a particular circuit depends only upon the nature of the external resistors and capacitors. In practice, there is a limit to the open-loop voltage gain of an op-amp which limits the upper frequency that may be used.

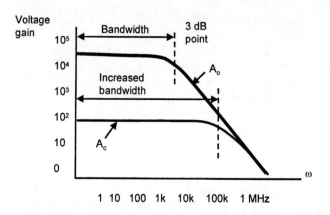

The upper frequency limit is due to the presence of internal capacitances within the IC itself which are present intentionally to enhance stability under feedback conditions.

The bandwidth increases with decreasing voltage gain (increasing negative feedback).

For a 741 IC, the (gain × bandwidth) product is fairly constant at about 1 MHz. The roll-off is about 25 dB/decade.

3.3.9 Review questions

1. The difference amplifier shown has a gain of 10 and a CMRR of 60 dB.

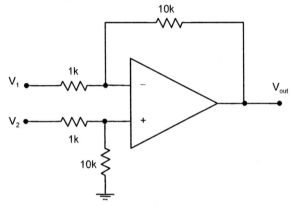

(a) What are the disadvantages of using this circuit as an instrumentation amplifier?

(b) Design a cross-coupled input stage for this amplifier to provide an overall gain of 100 and calculate the new CMRR.

(c) If this modified amplifier were presented with a 50 mV difference signal with 20 mV of common mode noise, determine the nature of the output voltage.

2. A logarithmic amplifier is constructed using a diode whose reverse bias leakage current I_o is 200 nA. At room temperature, the following relation applies to the diode:

$$I \approx I_o e^{eV/kT}$$

where $e/kT = 40$ at $T = 300$ K. The diode has a maximum forward bias current rating of 50 mA. In the circuit below, determine a suitable value of R_1 to provide an output of 0.5 V for an input of 20 V.

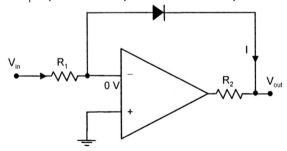

3. A chromel–alumel thermocouple is being used to measure temperature.
   A voltage appears at the thermocouple outputs which is dependent of
   the temperature difference between the cold and hot junctions. The
   voltage is typically a few millivolts. A digital output display shows the
   temperature in °C and is driven by an analog to digital converter. The
   ADC converts a signal from 0 to 10 V to an 8-bit digital value. The
   maximum temperature to be measured is 1000 °C.

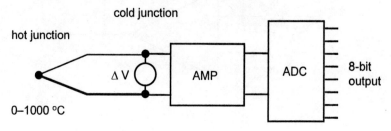

(a) Design an instrumentation amplifier which converts the output
    from the thermocouple to that required to utilise the full input
    range of the ADC.

(b) Determine the resolution of the system.

4. Under what circumstances is it better to use a difference amplifier than
   a single-ended input amplifier?

3.3.10 Activities

Instrumentation amplifier

1. Construct a simple difference amplifier and measure the difference gain and CMRR when the amplifier is driven by high impedance sources (you may have to connect a large resistor in series with the input signal). Comment on the results.

2. Add unity gain voltage followers to each input and measure the difference gain and CMRR of the amplifier. Compare with previous measurements and comment. (Note: A better op-amp than a 741 may be required – e.g. LM308N.)

Difference amplifier

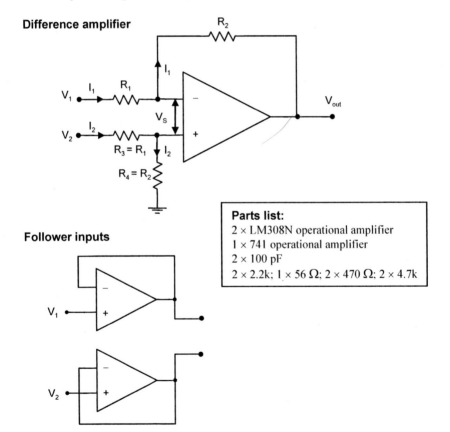

Follower inputs

Parts list:
2 × LM308N operational amplifier
1 × 741 operational amplifier
2 × 100 pF
2 × 2.2k; 1 × 56 Ω; 2 × 470 Ω; 2 × 4.7k

3. Design and construct a cross-coupled input stage for a difference amplifier but do not connect to the amplifier yet. The cross-coupled input stage is to have a gain of about 100.

4. Measure the gain of the input stage for various values of R.

5. Measure the common mode gain of the input stage and comment.

6. Connect to the difference amplifier and measure the gain of the overall circuit, its common mode rejection ratio, and bandwidth. Comment on the results.

7. Suggest how the gain of the amplifier should be distributed over the two stages for optimum performance.

8. Connect the output of the thermocouple circuit from Part 1 of this book to the instrumentation amplifier input, and connect the instrumentation amplifier output to the analog input of the data acquisition system from Part 2 of this book.

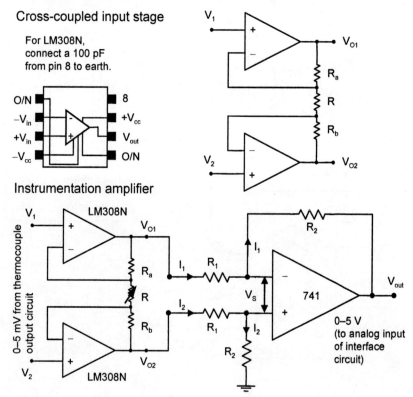

Cross-coupled input stage

For LM308N, connect a 100 pF from pin 8 to earth.

Instrumentation amplifier

Log amplifier

The principal problem with log amps is drift due to the input bias current of the op-amp. An FET op-amp is usually used to minimise this error. In the circuit below, an LF356 FET input op-amp is used.

The bias current is roughly balanced with R_1 (on the positive input) of about 100 kΩ. A capacitor 0.01 µF across the transistor helps to stabilise the circuit against high frequency (RF) oscillations. A large resistor (20 MΩ) is needed in parallel to keep output drift low.

For offset voltage balancing, a multi-turn pot is required since the input voltage has to span a wide range.

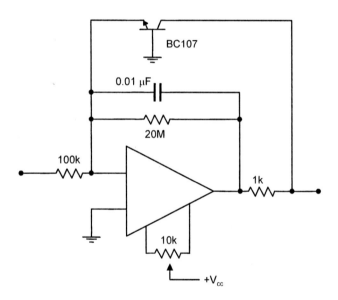

1. Connect the op-amp as an inverting amplifier with a gain of about 100 (i.e. do not use the transistor as a feedback element yet). With $V_{in} = 0$, balance the offset voltage to zero.

2. Now connect the transistor and stabilising components as shown above.

3. Vary V_{in} over a wide range to get a rough idea if the circuit is working. V_{out} will not change very much (due to it being a log amplifier).

4. When you are satisfied that the circuit is working, vary V_{in} from about 90 V down to about 10^{-3} V with about four points per decade.

5. Plot the gain of the circuit with V_{out} on the vertical axis and V_{in} on the horizontal log axis. If you have only linear paper available, plot V_{out} vs $\log V_{in}$.

6. Estimate values for the transfer function from your experimental readings and compare with calculated values.

7. If time permits, try making a log amplifier using a 741 op-amp and record your findings.

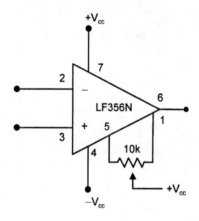

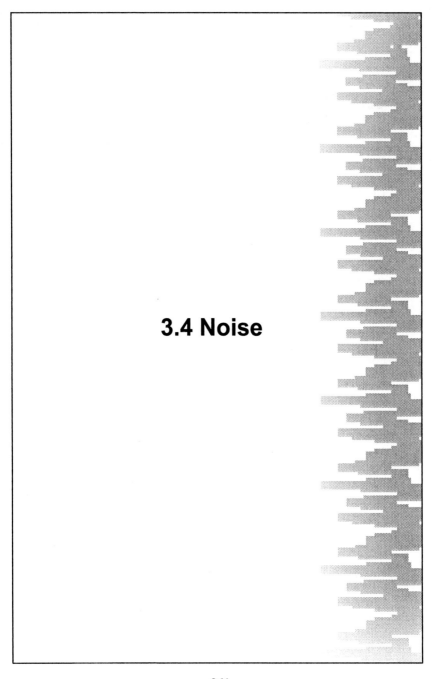

3.4 Noise

3.4.1 Intrinsic noise

Thermal (Johnson or Nyquist*) noise

* Johnson did the experiments, Nyquist developed the equation.

$$V_{n\,rms} = \left[4kTR\Delta f\right]^{\frac{1}{2}}$$

Power

k	Boltzmann's constant
	1.38×10^{-23} J/K
T	Absolute temperature
R	Resistance
Δf	Bandwidth

$$V_n{}^2 = 4kTR\Delta f$$

$$P = \frac{V^2}{R}$$

$$= 4kT\Delta f$$

e.g. 10k resistor at 300 K over a bandwidth of 10 kHz gives an rms noise figure of 1.3 μV

Noise power is proportional to T and Δf.

Thermal noise and shot noise are present at all frequencies and is called **white** noise. Noise may be reduced by reducing any terms in the expression, e.g. reducing the temperature, resistance and the bandwidth.

Strictly speaking, white noise is noise which has a constant power density at all frequencies over the band of frequencies of interest.

Shot noise

Associated with the randomness of charges moving across a potential barrier.

• thermoionic emission
• contact points

$$i_n{}^2 = 2eI_s\Delta f$$

Power

e	charge on electron 1.6×10^{-19} C
I_s	DC signal current (A)
i_n	noise current (A)
Δf	bandwidth (Hz)

$$P_n = i_n{}^2 R$$

$$= 2\Delta fei_n R$$

> Note: In general, noise increases with bandwidth. Noise is due to random fluctuations which can contribute to a significant high frequency component of the total signal. Reducing the bandwidth reduces the amount of high frequency noise.

Flicker noise

This type of noise increases with decreasing frequency and is sometimes called 1/f noise. For this reason, sensitive measurements should not be made using DC. The precise origin of flicker noise is not well understood. It is usually not important compared to other noise above 1 kHz.

3.4.2 Environmental noise

Types of environmental noise

This type of noise arises from sources outside the measuring system. Environmental noise is often called **interference**. Interference may be mechanical in nature (from mechanical **vibrations**) or electrical. Electromagnetic interference (**EMI**) is the most common. Such noise may arise from:

- Radiation from the abrupt cessation of electric current during the switching off or control of heavy machinery.
- Radiation from AC circuits such as power lines, rectifiers, etc.
- Lightning.

Noise from the above generally occurs at low frequencies. Noise can also occur at radio frequencies (RF). **RF noise** can arise from:

- Transmitters (two-way radios, cell phones, radar installations, etc.).
- Electronic devices working at high frequencies.

A third source of noise, or even damage, is **electrostatic discharge**. This is very prominent in dry weather and depends on the material used for the equipment and furnishing the surroundings.

Reduction of environmental noise

The best method of reducing the effects of noise is to reduce the noise at its source. This is not always practical, so the next best method is to attempt to divert the noise signal to ground through the use of filters before it is registered by the transducer. Failing that, the most common approach for reducing EMI is to be careful with the physical location of sensitive components. The major contribution to the effect of noise occurs at the first stage of amplification. For this reason, a **preamplifier** should be located as close to the transducer as possible. The preamplifier should be a differential amplifier with a good CMRR. Signal leads from the pre-amp to the main power amplifier should be shielded cable and be routed away from transformers and mechanical switches. All shields should be grounded at a common point so as to eliminate ground loops.

3.4.3 Signal-to-noise ratio

A measure of the relative magnitude of the noise is usually given by the **signal-to-noise ratio**, or SNR (often expressed in dB).

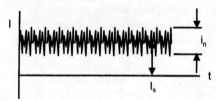

The signal-to-noise ratio is the ratio of the signal power over the noise power.

$$SNR = \frac{P_s}{P_n}$$

The larger the SNR the better.

$$SNR_{db} = 10 \log_{10} \frac{P_s}{P_n}$$

Power is proportional to I^2 (or V^2), hence:

$$SNR_{db} = 20 \log_{10} \left| \frac{V_s}{V_n} \right| \quad \longleftarrow \quad \text{rms values usually used}$$

Noise is always present in the original signal and may be amplified and new noise added by the instrumentation amplifier itself. The degradation of SNR from the input to the output of an amplifier is called the "**Noise Figure**" NF. An NF of less than about 3 dB is considered good.

$$NF = \frac{SNR_s}{SNR_o} \quad \begin{matrix} \longleftarrow \text{Signal} \\ \\ \longleftarrow \text{Amplifier output} \end{matrix}$$

Noise in a transistor (such as a BC109 PNP BJT) arises from **thermal noise** from the resistance of the semiconductor itself and shot noise from the passage of charge carriers across the pn junctions. **Flicker noise** is also present and is due to the randomness of the diffusion process of carriers. Flicker noise, being 1/f dependent, is the main source of noise at low frequencies (<1 kHz) in a transistor. In an FET, shot noise is not so important since the pn junction is in reverse bias and the gate current is very small.

3.4.4 Optical detectors

Consider the factors that affect the rate of electron production within an optical detector:

$$r = \eta \frac{P_s}{h\nu}$$

r	rate of electron production (electrons/second)
η	quantum efficiency
h	Planck's constant (6.63×10^{-34})
ν	Frequency in Hz
P_s	Incident power (i.e. the signal)

Now, the resulting **signal current** I_s is simply:

$$I_s = re$$

e   charge on electron

$$= \eta \frac{eP_s}{h\nu}$$

The **noise current** is found from:

$$i_n^2 = 2\Delta f e I_s$$

Δf is called the **bandwidth**.

$$= 2\Delta f e^2 \frac{\eta P_s}{h\nu}$$

This is noise from "internal" sources in the detector

The signal to noise ratio is thus:

$$SNR = \eta^2 \frac{e^2 Ps^2}{h^2\nu^2} \frac{h\nu}{2\Delta f e^2 \eta P_s} = \frac{P_s}{P_n}$$

Note:

$$P = I^2 R$$

$$\therefore SNR = \frac{I_s^2}{i_n^2}$$

$$\frac{P_s}{P_n} = \frac{\eta P_s}{2h\nu\Delta f}$$

$$P_{smin} = \frac{2h\nu\Delta f}{\eta}$$ NEP$_{signal\ limited}$ (proportional to Δf)

The minimum detectable signal occurs when SNR = 1, that is, when $P_{smin} = P_n$. This is called the **Noise Equivalent Power** NEP$_{signal\ limited}$.

If the background radiation noise power P_b is $>>$ than signal power P_s, then the signal is background limited and the SNR becomes:

$$SNR = \frac{P_s}{P_n} = \frac{\eta P_s^2}{(P_s + P_b)2h\nu\Delta f}$$

$$= \frac{\eta P_s^2}{P_b 2h\nu\Delta f}$$

This is noise from background radiation which is incident on the detector along with the signal.

$$P_{smin} = \sqrt{\frac{P_b 2h\nu\Delta f}{\eta}}$$

This is the NEP$_{background\ limited}$ (proportional to $\sqrt{\Delta f}$)

3.4.5 Lock-in amplifier

The **lock-in amplifier** uses a phase detection circuit where the amplitude of the output signal is proportional to the amplitude of the input signal and proportional to the cosine of the phase difference between the input signal and a reference signal – the input signal and reference signal must have the same frequency.

The lock-in amplifier thus requires a reference signal, and a periodic input signal. A slowly varying input signal, which contains noise, can be made periodic, or repetitive, by **chopping**.

> For example, a particular input signal could consist of a slowly varying DC voltage (e.g. thermocouple output). By "chopping" the signal the DC output is converted into a square wave of known frequency.

The phase detector produces an output signal which follows variations in the amplitude of the input signal (if the frequency of the input and reference signals are the same – which is ensured by having the reference signal also operate the chopper) and where the phase of the reference signal is the same as that of the signal.

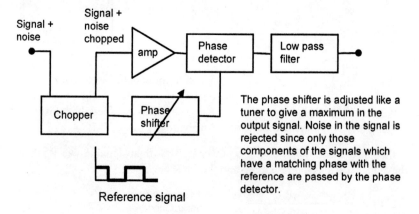

The phase shifter is adjusted like a tuner to give a maximum in the output signal. Noise in the signal is rejected since only those components of the signals which have a matching phase with the reference are passed by the phase detector.

3.4.6 Correlation

This technique can be performed electronically or numerically on the recorded data. The signal of interest must be repetitive. There are two general types of correlation:

- **cross-correlation**
- **auto-correlation**

Both methods employ a **reference signal** (the auto-correlation uses the sample signal, shifted in time, as the reference). During a time period, the reference signal is delayed by a delay τ. The sample signal and the delayed reference signal are multiplied together and then added. The reference is shifted again and multiplied and added to the original signal, and so on until a complete period has been **correlated**.

Mathematically, the correlated output V_{out} is given by:

$$V_{out} = \frac{1}{T} \int_{-T}^{T} V_{in}(t) V_{ref}(t - \tau) dt$$

where for auto-correlation, $V_{ref} = V_{in}$.

Auto-correlation allows us to indirectly obtain information about the frequencies present in a signal but not necessarily the waveform of that component. Cross-correlation gives information about the waveform of the signal of interest.

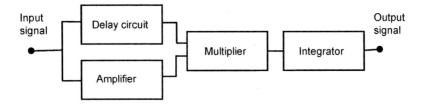

3.4.7 Review questions

1. Calculate the open circuit rms noise voltage over the frequency range 0–1 MHz between the terminals of a 100 kΩ resistor at a temperature of 27 °C. Also calculate the available noise power from this resistor.
   (Ans: 0.04 mV, 4.14 × 10⁻¹⁵ W)

2. A transducer of resistance 1 MΩ delivers a signal current of 10 μA to an amplifier which has an input resistance of 10 MΩ and an bandwidth of 10 kHz. Calculate the temperature to which the transducer must be cooled to achieve an SNR less than 130 dB.   (Ans: −73 °C)

3. A photodiode provides a signal current of 100 nA under a constant level of illumination. What is the rms shot noise current in the diode over a bandwidth of 100 kHz?   (Ans: 5.66 × 10⁻¹¹ A)

4. A signal of interest occurs at a frequency of 300 Hz and provides a voltage at the transducer output of 100 mV. A combination of thermal and flicker noise is present which is approximately described by:

   $$V(f) = a + \frac{b}{f} \text{ volts/Hz } \quad a = 2.5 \times 10^{-6}; b = 4.0 \times 10^{-3} \quad \text{(Ans: 35 db)}$$

   Calculate the SNR if the output is filtered to within 250–350 Hz.

5. An optical detector is required to have 15 dB signal to noise power ratio. The noise equivalent power (NEP) is background limited.

   (a) Calculate the power ratio P_s/P_n. Given that the response of the detector is limited by the background noise, what is the minimum detectable power expressed in terms of the NEP.

   (b) Find the ratio of the minimum detectable power P_{smin} to the background power P_B at 300 K and $\lambda = 8$ μm if a temperature resolution of $\Delta T = 0.05$ °C is required.

   $P_{smin} = \Delta P = $ power resolution

   $$\frac{dP}{dt} = \frac{P}{T}\left(\frac{h\nu}{kT}\right) \text{ from Planck's equation}$$

   (c) Express $\Delta P/P_B$ in terms of P_B, ν and Δf. Hence find an expression for P_B in terms of Δf.

   (d) Calculate the product of detector aperture area A and the bandpass $\Delta\nu$ required for $\Delta f = 5$ MHz for a field of view of solid angle $\Omega = 10^{-6}$ Sr given that: $P\Delta f = \dfrac{A\Omega 2h\nu^3 e^{-h\nu/kT}}{c^2}\Delta\nu$

   (e) Estimate the aperture A required of the detector for $\Delta\nu = 10^{12}$ Hz.

   (Ans: 31.6 NEP, 0.001, 5 × 10⁻¹⁴, 0.128 m²)

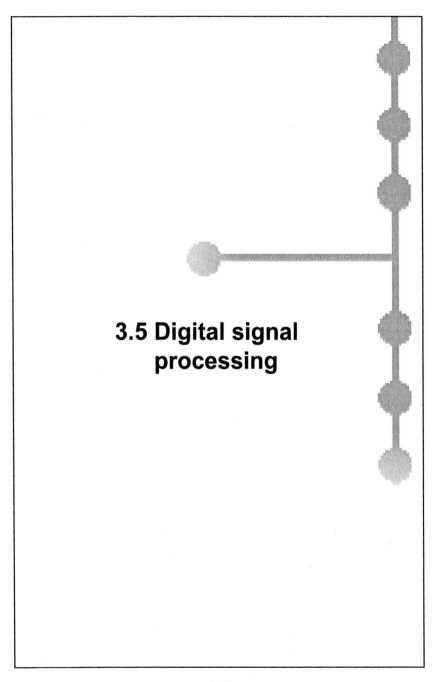

3.5 Digital signal processing

3.5.1 Digital filters

If a continuous signal y(t) is sampled N times at equal time intervals Δt, then the resulting digitised signal includes the information of interest plus any noise that might have been present in the original signal.

The purpose of a digital filter is to take this set if data, perform mathematical operations on it, and produce another set of data possessing certain desirable properties (such as reduced noise).

Digital filters fall into two basic categories: **Infinite Impulse Response (IIR)** or **Finite Impulse Response (FIR)**. These terms describe the time domain characteristics of the filter when presented with an impulse signal as an input.

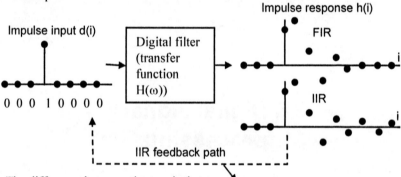

The difference between the two is that for the FIR, the output of the filter decays to zero for an impulse input. The output of an FIR filter depends on the present, and previous inputs (the filter is non-recursive). The output from an IIR filter employs a feedback mechanism so that the output depends upon past outputs as well as past and present inputs (i.e. recursive).

The feedback associated with the IIR filter presents an input to the filter even when the external input drops to zero. The output never reaches zero. It may decrease, oscillate, or even become unstable and increase without limit. IIR filters are usually more computationally efficient than their FIR counterparts.

There are two approaches to digital filtering. The data itself may be operated upon using a filter algorithm with the desired transfer function, or, the frequency spectrum of the data may be obtained using **Fourier analysis**, selected frequencies discarded, and then the filtered sequence recomputed from the modified spectrum. The second method is described in some detail in this chapter.

3.5.2 Fourier series

The **Fourier series** gives amplitudes and frequencies of the component sine waves for any periodic function f(t). For periodic functions of period T_0 with frequency ω_0, the Fourier series can be written:

$$f(t) = A_0 + \sum_{n=1}^{\infty} \left[A_n \cos n\omega_0 t + B_n \sin n\omega_0 t \right]$$

with

$$A_0 = \frac{1}{T_0} \int_0^{T_0} f(t)\,dt; \quad A_n = \frac{2}{T_0} \int_0^{T_0} f(t)\cos n\omega_0 t\,dt; \quad B_n = \frac{2}{T_0} \int_0^{T_0} f(t)\sin n\omega_0 t\,dt$$

"DC" term or average value of f(t)

Amplitude terms for component frequency $n\omega_0$

Using **Euler's formula**, it can be shown that any cosine (or sine) function can be represented by a pair of exponential functions:

$$\cos \omega t = \frac{1}{2}\left[e^{j\omega t} + e^{-j\omega t}\right]; \quad \sin \omega t = -j\frac{1}{2}\left[e^{j\omega t} - e^{-j\omega t}\right]$$

Substituting into the Fourier series we obtain:

$$f(t) = \sum_{n=-\infty}^{\infty} C_n e^{nj\omega_0 t}$$

$$C_n = \frac{1}{T_0} \int_0^{T_0} f(t) e^{-nj\omega_0 t}\,dt$$

Note: C_n is a complex number, the real part contains the amplitude of the cos terms, and the imaginary part the amplitude of the sin terms.

$$C_n = \frac{1}{2}\left(A_n - jB_n\right) \quad n > 0$$

$$= \frac{1}{2}\left(A_n + jB_n\right) \quad n < 0$$

A plot of C_n vs frequency is a frequency spectrum of the signal. For example, if f(t) =A cos ω_0t, then the frequency spectrum is a *pair* of lines of height A/2 located at $\pm\omega_0$.

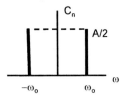

Negative frequencies arise from the representation of sinusoidal signals by a pair of exponential functions. There is no "DC" term since the average value of the function is zero.

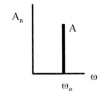

This plot is the magnitude of the *exponential* components of the signal. A frequency spectrum using trigonometric coefficients would be a single line of height A at ω_0.

3.5.3 Fourier transform

Non-periodic functions are functions with an infinite period, i.e. $T_0 \rightarrow \infty$ and $\omega_0 \rightarrow 0$. The component frequencies can no longer be represented as discrete spectral lines, but take on an infinitely continuous range of values. Now, for the periodic case, we have:

$$C_n = \frac{1}{T_0} \int_0^{T_0} f(t) e^{-nj\omega_0 t} dt$$

As $T_0 \rightarrow \infty$, $C_n \rightarrow 0$. That is, the amplitude of the spectral lines becomes vanishingly small as the spectral lines merge into a continuum. But, the integral is finite, hence, the product $C_n T_0$ can be written:

The product $n\omega_0$ becomes the continuous variable ω, hence:

$$C_n T_0 = \int_{-\infty}^{\infty} f(t) e^{-nj\omega_0 t} dt$$

$$C_n T_0 = \int_{-\infty}^{\infty} f(t) e^{-j\omega t} dt$$

$$= F(\omega)$$

or: $C_n = \dfrac{F(\omega)}{T_0} = \dfrac{\omega_0}{2\pi} F(\omega)$

Now, going back to the periodic case, and replacing $n\omega_0$ with the continuous variable ω, we obtain:

$\longrightarrow \quad f(t) = \displaystyle\sum_{n=-\infty}^{\infty} C_n e^{nj\omega_0 t}$

$$f(t) = \sum_{n=-\infty}^{\infty} \frac{\omega_0}{2\pi} F(\omega) e^{j\omega t} = \frac{1}{2\pi} \int_{-\infty}^{\infty} F(\omega) e^{j\omega t} d\omega \quad \substack{\text{Fourier} \\ \text{integral}}$$
$$\omega_0 \rightarrow 0$$

The function $F(\omega)$ is called the **Fourier transform** of $f(t)$ and is written $F[f(t)]$. The Fourier transform is a continuous function of ω.

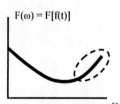

$F(\omega) = F[f(t)]$

Indicates possible high frequency noise in the original signal

The function $f(t)$ is often termed the **inverse Fourier transform** of $F(\omega)$: $f(t) = F^{-1}[F(\omega)]$

Because of the continuous nature of ω, the amplitude of the frequency component of the signal at any one particular frequency approaches zero. $F(\omega)$ is really a frequency density function: $F(\omega) = C_n T_0 = C_n \dfrac{2\pi}{\omega_0}$

Thus, it is more appropriate to say that the frequency spectrum of a signal has an amplitude of say 5 V per rad at a particular value of ω.

3.5.4 Sampling

If a continuous signal y(t) is sampled N times at equal time intervals Δt, then the **sampling frequency** is: $\omega_s = 2\pi/\Delta t$.
The digital signal has a duration $T_o = N\Delta t$.

Now, a non-periodic sample sequence of finite length N sampled over a time T_o can be considered to be periodic, for the purposes of analysis, with a period T_o. For a periodic signal, the frequency spectrum of the signal consists of lines spaced $\omega_o = 2\pi/T_o$ apart where ω_o is the fundamental frequency of the signal.

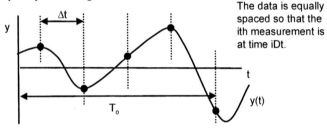

The data is equally spaced so that the ith measurement is at time iDt.

Thus, we can say that the continuous frequency spectrum of a **non-periodic** signal can be completely specified by a set of regularly spaced frequencies a minimum of $\Delta\omega = 2\pi/T_o$ radians/sec apart. That is, for the purposes of analysis, we say that a **non-periodic signal of finite length** is periodic with a period equal to the length of the signal. The component frequencies consist of equally spaced intervals:

$$\Delta\omega = \frac{2\pi}{T_o} = \frac{2\pi}{N\Delta t} = \frac{\omega_s}{N}$$

How *many* component frequencies are required to represent the data? Well, the spectrum of a digital signal is always periodic with the same set of frequencies repeating over and over with a frequency period of $2\pi/\Delta t$. If we sample at intervals of $\Delta\omega = 2\pi/N\Delta t$, then the total number of frequencies per period is just

$$\frac{2\pi}{\Delta t}\frac{N\Delta t}{2\pi} = N$$

The frequency components contained within the N data points are:

$$\omega_k = \frac{2\pi k}{N\Delta t} = \frac{k}{N}\omega_s \quad \text{where k goes from 0 to N} - 1$$

The **frequency resolution** $\Delta\omega$ is ω_s/N. The greater the value for N, the finer the resolution of the frequency bins or **channels** used to represent the original signal.

3.5.5 Discrete Fourier transform

Now, in general, $F(\omega)$ is given by the integral:

$$F(\omega) = \int_{-\infty}^{\infty} f(t)e^{-j\omega t}dt$$

We can approximate this to a finite sum as:

$$F(\omega) = \sum_{i=0}^{N-1} y_i(i\Delta t)e^{-j\omega(i\Delta t)}\Delta t$$

where N is the total number of equally spaced data points and $y_i(i\Delta t)$ is the actual data at i recorded at time $i\Delta t$. But, ω in this formula is a continuous variable; however, for a discrete number of samples:

$$\omega_k = \frac{2\pi k}{N\Delta t}$$

Further, recalling that $F(\omega)$ is actually a frequency density function, we can thus write the actual **amplitude spectrum** of the signal as:

$$\frac{F(\omega_k)}{\Delta t} = \sum_{i=0}^{N-1} y_i(i\Delta t)e^{-j2\pi ik/N}$$
$$= C(k)$$

where k goes from 0 to N − 1

In terms of sines and cosines, we have:

$$C(k) = \sum_{i=0}^{N-1} y_i(i\Delta t)\left[\cos\frac{2\pi ik}{N} - j\sin\frac{2\pi ik}{N}\right]$$

This formula can be easily implemented in software

> **Fast Fourier transform** FFT
> Computation of the DFT is time consuming, requiring in the order of N^2 floating-point multiplications. However, many of the multiplications are repeated as i and k vary. The FFT is a collection of routines which are designed to reduce the amount of redundant calculations. Each different implementation of the FFT contains different features and advantages. Most pre-written computer subroutines employ some sort of FFT routine to find a DFT of a series of samples. The algorithm used in some computer languages is known as the "split-radix" algorithm and requires approximately $N\log_2 N$ operations.

Each value of C(k) is a complex number of the form A − Bj. The complete array of N complex numbers comprising this series is called a **discrete Fourier transform** or DFT. Note that C(k) is periodic (with period $2\pi/\Delta t$). If we perform the sum past k = N − 1, we get the same set of frequency components again and again.

3.5.6 Filtering

Consider a signal $y_{in}(t)$ which contains high frequency noise:

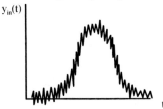

The high frequency noise would lead to an increase in $F[y_{in}(t)]$ at high frequencies. If it were possible to separate out the signal of interest y_s from the noise y_n, $F[y(t)]$ for each would be something like:

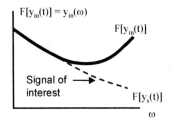

The signal can be represented by a **Fourier integral**:

$$y_{in}(t) = \int_{-\infty}^{\infty} V_{in}(\omega) e^{j\omega t} d\omega$$

Fourier transform in ω-domain

Now, consider the transfer function $H(\omega) = y_{out}/y_{in}$ of an ideal low pass filter in the ω-domain.

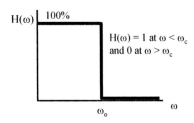

$H(\omega) = 1$ at $\omega < \omega_c$ and 0 at $\omega > \omega_c$

If this function $H(\omega)$ were multiplied with the Fourier transform of y_{in}, then the amplitude of all components of the Fourier transform above ω_0 would be reduced to zero thus eliminating (or at least reducing) the high frequency noise component of the signal.

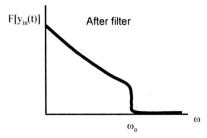

The resulting Fourier series or integral (using the modified transform as the amplitude coefficients) then represents the **filtered signal** in which noise is reduced.

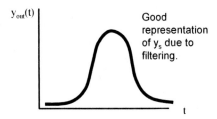

Good representation of y_s due to filtering.

3.5.7 Digital filtering (ω-domain)

For a given input signal $y_{in}(t)$, the filtered signal $y_{out}(t)$ is obtained from the inverse Fourier transform of the product $H(\omega)F(\omega)$.

$$y_{out}(t) = \frac{1}{2\pi}\int_{-\infty}^{\infty}H(\omega)F(\omega)e^{j\omega t}d\omega$$

$$F(\omega) = \int_{-\infty}^{\infty}y_{in}(t)e^{-j\omega t}dt$$

For an array of **discrete samples**, we must replace the continuous variable ω with:

$$H(\omega) = 1 \qquad \omega \leq \omega_c$$
$$= 0 \qquad \omega > \omega_c$$

$$\omega_k = \frac{2\pi k}{N\Delta t}$$

The **discrete Fourier transform** of y_i is:
$$\frac{F(\omega_k)}{\Delta t} = \sum_{i=0}^{N-1}y_i(i\Delta t)e^{-j2\pi ik/N}$$
$$= C(k)$$

where k goes from 0 to N – 1

The **inverse transform** is found from:

$$y_i(i\Delta t) = \frac{1}{2\pi}\int_{-\infty}^{\infty}F(\omega_k)e^{j\omega i\Delta t}d\omega = \frac{1}{N}\sum_{k=0}^{N-1}C(k)e^{j2\pi ik/N}$$

or

complex numbers

$$y_i(i\Delta t) = \frac{1}{N}\sum_{k=0}^{N-1}C(k)[\cos 2\pi ik/N + j\sin 2\pi ik/N]$$

Thus, the filtered sequence is given by:

$$y_{out}(i) = \frac{1}{N}\sum_{k=0}^{N-1}H(\omega_k)C(k)e^{j2\pi ik/N}$$

Note: The product $H(\omega)C(k)$ involves complex numbers.

As written here, $y_{out}(i)$ is an array of complex numbers containing both phase and magnitude information. The magnitudes may be obtained by taking the square root of the sum of the squares of the real and imaginary values.

Now, since

$$\omega_k = \frac{2\pi k}{N\Delta t} = \frac{k}{N}\omega_s$$

Then the filter transfer function can be expressed in terms of the parameter k/N which goes from 0 to 1.

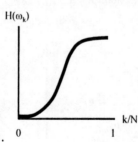

$H(\omega_k)$

k/N

0   1

Note: The DFT and inverse DFT allow for processing of complex signals (such as that in an AC circuit). If the data sampled is real, the imaginary terms in the reconstituted signal from the inverse transform reduce to zero.

3.5.8 Convolution

The filter transfer function H in the
$\omega-$ or s-domains is in fact a Fourier
transform! Although it does not show
the amplitude of the component sine
and cosine functions of a particular
signal, it does show the amplitude of
the ratio y_{out}/y_{in} against frequency for
the filter circuit.

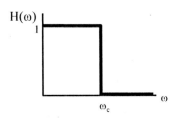

For a given input signal $y_{in}(t)$, the filtered signal $y_{out}(t)$ is obtained by the
Fourier integral given by:

$$y_{out}(t) = \int_{-\infty}^{\infty} H(\omega)F(\omega)e^{j\omega t}d\omega$$

$y_{out}(t)$ is the inverse Fourier transform
of the product $H(\omega)F(\omega)$.

$$F(\omega) = \frac{1}{\sqrt{2\pi}} \int_{-\infty}^{\infty} y_{in}(t)e^{-j\omega t}dt$$

$$H(\omega) = 1 \qquad \omega \leq \omega_c$$
$$= 0 \qquad \omega > \omega_c$$

If the functions h(t) and $y_{in}(t)$ are the original functions in the time domain,
we say that the output signal $y_{out}(t)$ is the **convolution** of these two
functions.

$$y_{out}(t) = \int_{-\infty}^{\infty} F[y_{in}(t) * h(t)]e^{j\omega t}d\omega$$

$$F[y_{in}(t) * h(t)] = F(\omega)H(\omega)$$
$$\uparrow \qquad = F[y_{in}(t)]F[h(t)]$$
convolution

A convolution is a special type of
superposition in the time domain. It is
a weighted sum of products of two
signals. It is equivalent to a
multiplication in the frequency domain.

Digital filtering can be done in
either the time (using convolutions)
or frequency (using Fourier
transforms) domains.

> **Modulation** is another type of
> superposition. Consider the
> formula:
> $$y(t) = Ae^{j\omega t}$$
> The frequency term is modulated
> by the amplitude term which may
> change as a function of time. This
> is called amplitude modulation.
> Modulation is a multiplication of two
> signals in the time domain. A
> multiplication in the time domain is
> equivalent to a convolution in the
> frequency domain.

3.5.9 Discrete convolution

To find the convolution of y(i) and h(i) we multiply the individual impulse with the impulse response function suitably shifted so as to align with the impulse. Then, all these are added together to obtain the final filtered digital signal.

A digital signal can be thought of as a summation of a series of impulses, each offset by a sampling interval.

Let the impulse response of a particular digital filter be represented by h(i):

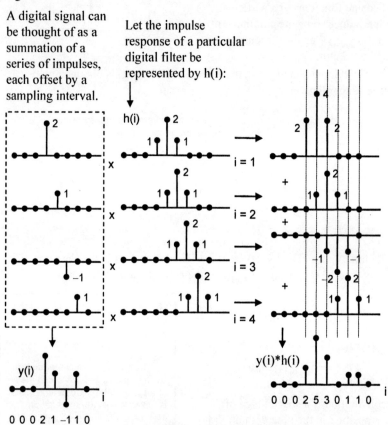

y(i)

0 0 0 2 1 −1 1 0

Digital signal is sum of shifted impulses

y(i)*h(i)

0 0 0 2 5 3 0 1 1 0

In general, a **discrete convolution** between two arrays of discrete samples $y_{in}(i)$ and h(i) is given by:

$$y_{out}(n) = \sum_{x=0}^{n} y_{in}(x) h(n-x)$$

n goes from 0 to $(N_y + N_h - 1)$. Arrays are numbered from 0 to N − 1 and contain N elements.

3.5.10 Digital filtering (t-domain)

The **time domain** approach is often used in digital filtering in preference to the ω-domain approach. There are several methods of approximating the desired ideal filter transfer function. The transfer functions of two popular low pass methods are shown:

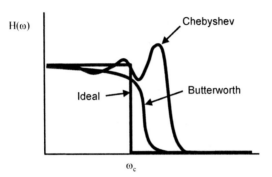

The actual filtering function is an equation (called a **difference equation**) which gives the smoothed value of y_i in terms of the actual value of y_i and various weighted combinations of previous and future values of y.

A simple (non-recursive) difference equation which performs a five point weighted averaging is:

$$\hat{y}_i = a_0 y_i + a_1\left(y_{i+1} + y_{i-1}\right) + a_2\left(y_{i+2} + y_{i-2}\right)$$

The **impulse response** of a particular difference equation is found by setting y_i to 1 when $i = 0$ and y_i to 0 otherwise. For example:

$$\hat{y}_0 = a_0 1 + a_1\left(0+0\right) + a_2\left(0+0\right) = a_0$$
$$\hat{y}_1 = a_0 0 + a_1\left(0+1\right) + a_2\left(0+0\right) = a_1$$
$$\hat{y}_2 = a_0 0 + a_1\left(0+0\right) + a_2\left(0+1\right) = a_3$$

The procedure for finding the form of the difference equation and the values of the coefficients for a particular digital filter is for an advanced course in DSP although we will look at a simple example next.

A computer program can be used to calculate the following sum to obtain the filtered value of y_i.

$$y_i = \frac{1}{b_0}\left[\sum_{j=0}^{N-1} a_j x_{i-j} - \sum_{k=1}^{M-1} b_k y_{i-k}\right]$$

N is the number of forward coefficients (a_j) and M the number of reverse coefficients (b_k).

3.5.11 Example

Non-recursive moving average filter

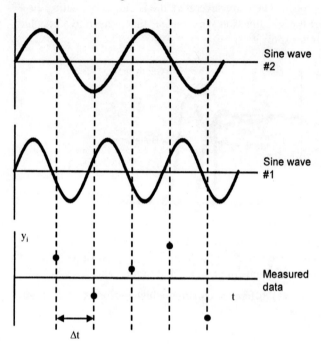

Consider a set of measurements of a quantity y taken at equal time intervals Δt (a sampling frequency of $1/f$).

Let the measured value y_i at $i\Delta t$ be expressed as a sum of two sine waves (plus a DC term if needed @$k = 0$) making $N = 3$.

$$y_i(i\Delta t) = \frac{1}{3}\sum_{k=0}^{2} F(\omega) e^{j2\pi ik/N}$$

Note: The data is equally spaced so that the ith measurement is at time $i\Delta t$.

We require an equation which gives the best estimated values of y_i at any particular value of i. That is, a smoothing equation.

Let the smoothed value of y_i at each data point be given by the formula:

$$\hat{y}_i = a_0 y_i + a_1(y_{i+1} + y_{i-1}) + a_2(y_{i+2} + y_{i-2})$$

This is a five point smoothing equation (the **difference equation**).

3.5.12 Smoothing transfer function

y_i are the actual data points and y_i are the smoothed or fitted data points. The constants a_0, a_1 and a_2 are to be chosen so as to smooth any noise from the actual data. Contributions to the smoothed value of a particular data point come from the actual data and four neighbouring data points y_{i+1}, y_{i-1}, y_{i+2} and y_{i-2} making a five point **smoothing formula**. Now, each of the measured data points $\hat{y}_i$ can be calculated from the summation of the sine waves #1 and #2 evaluated at n = i. Substituting gives:

$$\hat{y}_i = a_0 y_i + a_1(y_{i+1} + y_{i-1}) + a_2(y_{i+2} + y_{i-2}) \quad \begin{array}{l}\text{Difference}\\\text{equation}\end{array}$$

$$\hat{y}_i = \sum_{k=0}^{2} F(\omega_k) \left[a_0 e^{j2\pi ik/N} + a_1\left(e^{j2\pi(i+1)k/N} + e^{j2\pi(i-1)k/N}\right) + a_2\left(e^{j2\pi(i+2)k/N} + e^{j2\pi(i-2)k/N}\right) \right]$$

$$= \sum_{k=0}^{2} F(\omega_k) e^{j2\pi ik/N} \left[a_0 + a_1\left(e^{j2\pi k/N} + e^{-j2\pi k/N}\right) + a_2\left(e^{2j2\pi k/Nt} + e^{-2j2\pi k/N}\right) \right]$$

$$= \sum_{k=0}^{2} F(\omega_k) e^{j2\pi ik/N} \left(a_0 + 2a_1 \cos 2\pi k/N + 2a_2 \cos 4\pi k/N \right)$$

but $e^{j\omega t} + e^{-j\omega t} = 2\cos\omega t$

$$= \sum_{k=0}^{2} F(\omega_k) H(\omega_k) e^{j2\pi ik/N}$$

where $\omega_k \Delta t = \dfrac{2\pi k}{N}$

$H(\omega)$ is a filter transfer function. Note that the form of $F(\omega)$ need not be known precisely since once the coefficients a_0, a_1 and a_2 have been decided, we can obtain y_i from the difference equation. If it is known that the original data contains say high frequency noise, then values for the smoothing coefficients are selected such that $H(\omega)$ has the form of a low pass filter. The coefficients a_0, a_1 and a_2 are found by simultaneous equations of data points taken directly from the desired transfer function. Three coefficients require three such data points ($H(\omega)$,k) to be used in the transfer function to obtain values of a_0, a_1 and a_2. These values are then used in the difference equation to give the fitted values of y_i at the desired value of i.

3.5.13 Review questions

1. Determine the real and imaginary parts of the DFT coefficients of the following "real" digital signals:

   (a) 1, 2, 3, 1, 2, 3
   (b) 0, 1, 1, 0

2. The data shows the average daily temperature in °C for each month of a year in Sydney, Australia.

   (a) Calculate the Fourier coefficients A_n, B_n for the following data for all the required values of k.
   (b) Draw a histogram of the data.
   (b) Draw a frequency spectrum.
   (c) Perform an inverse DFT to check your results of (a).

Jan	25
Feb	23
Mar	22
Apr	20
May	19
Jun	18
Jul	17
Aug	17
Sep	19
Oct	22
Nov	23
Dec	25

3. Using the method of a digital convolution, find $y_{out}(i) = y(i) * h(i)$ for the following number sequences:

   $$y(i) = 1, 3, 2, -1, 4 \qquad y_{out}(i) = \sum_{k=0}^{N-1} y_{in}(i) h(i-k)$$
   $$h(i) = 2, -1, 3$$

   (Ans: 2, 5, 4, 5, 15, −7, 12)

4. Design a seven point smoothing formula of the form:

   $$\hat{y}_i = a_0 y_i + a_1 (y_{i+1} + y_{i-1})$$
   $$+ a_2 (y_{i+2} + y_{i-2}) + a_3 (y_{i+3} + y_{i-3})$$

   which corresponds to a transfer function $H(\omega_k)$ approximating to the shape shown. Plot the actual transfer function calculated.

   Hint: Take H(0) = 0; H(1) = 1; H(0.5) = 0.5, H(0.8) = 1

   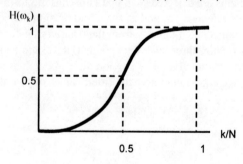

3.5.14 Activities

A temperature measurement system consists of a thermocouple connected to a difference amplifier, the output of which is fed into an ADC and then stored as an array of equally spaced readings on a microcomputer. Samples are taken every 5 seconds. Design a digital filter (in the language of your choice) which smoothes out any high frequency noise from this data. The smoothed data is to retain fluctuations which occur over a time greater than about a minute.

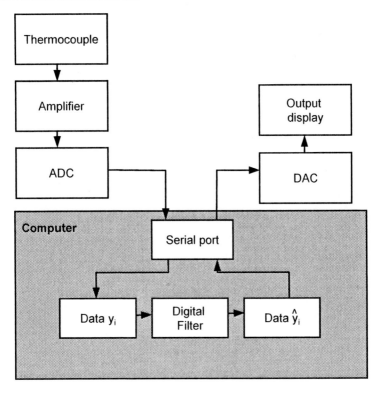

Index

Index

Further reading

R.M. Bertrand, "Programmable Controller Circuits ," International Thomson Publishing, 1995.

H.B. Boyle, D. Page, "Transducer Handbook: User's Directory of Electrical Transducers," Butterworth-Heinemann, 1999.

W. Buchanan, "Applied PC Interfacing: Graphics and Interrupts," Addison Wesley Longman, Inc., 1996.

F.M. Cady, "Microcontrollers and Microcomputers: Principles of software and hardware engineering," Oxford University Press, 1997.

D. Crecraft, S. Gergely, "Analog Electronics Circuits, Systems and Signal Processing," Butterworth-Heinemann, 2002.

A.J. Diefenderfer, B.E. Holton, "Principles of Electronic Instrumentation," 3rd Ed., International Thomson Publishing, 1994.

M. Elwenspoek, R. J. Wiegerink, "Mechanical Microsensors," Springer-Verlag NY, 2000.

D.R. Gillum, "Industrial Pressure, Level and Density Measurement," ISA, 1995.

A.R. Hambley, "Electrical Engineering: Principles and Applications," Prentice-Hall, Inc., 1997.

E.C. Ifeachor, B.W. Jervis, "Digital Signal Processing: A Practical Approach," 2nd Ed., Pearson Education, 2002.

J.H. Johnson, "Build your own low-cost data acquisition system and display devices," TAB Books, 1994.

M. Predko, "PC PhD: Inside PC Interfacing," McGraw-Hill Professional, 1999.

W.H. Rigby, T. Dalby, "Computer Interfacing: A Practical Approach to Data Acquisition and Control Lab Manual," Simon & Schuster, 1995.

I.R. Sinclair, "Sensors and Transducers," Butterworth-Heinemann, 2001.

G.A. Smith, "Computer Interfacing," Butterworth-Heinemann, 2000.

M.H. Tooley, "PC-based Instrumentation and Control," 3rd Ed., Butterworth-Heinemann, 2002.

W.A. Triebel, "The 80386, 80486, and Pentium Processors: Hardware, Software and Interfacing," Prentice-Hall, 1998.

M.J. Usher, D.A. Keating, "Sensors and Transducers," 2nd Ed., Macmillan Press Ltd, 1996.

R.M. White, R. Doering, "Electrical Engineering Uncovered," Prentice-Hall, 1997.

I. Busch-Vishniac, "Electromechanical Sensors and Actuators," Springer-Verlag, NY, 1998.

Parts lists for activities

Semiconductors

IC-CMOS/LSI UART	CDP6402CE
IC-COMS 8-BIT A/D	ADC0804LCN
IC-DUAL TRANSCEIVER	MAX232CPE
IC-74 SERIES TTL	DM7400N
IC-74HC SERIES CMOS	MC74HC04AN
IC-74HC SERIES CMOS	74HC393N
IC-74 SERIES TTL	DM7476N
IC-TEMPERATURE SENSOR	LM335H
IC-8-BIT DAC	DAC0800LCN
IC-SAMPLE AND HOLD	LF398N
IC-SUPER GAIN OP AMP	LM308H
IC-COMPENSATED OP AMP	UA741CN
DIODE,ZENER 4V7	BZX284-C4V7/T1
CRYSTAL,4.915200 MHZ	4079WT-4M9152
IC-FET OP AMP	LF356H

Part 1
1 × LM335H precision temperature reference
1 × IN4732 4.8V zener diode
2 × 220k; 1 × 2.2k; 1 × 100 Ω; 1 × 4.7k; 1 × 680 Ω; 1 × 1K

Part 2
1 × ADC0804 A to D converter
1 × 6402 UART
1 × 7400 NAND
1 × 232CPE RS232 line driver
1 × 74HC04 hex inverter CMOS
1 × 74HC393 CMOS counter
1 × 4.9152 MHz crystal
2 × 33 pF; 1 × 147 pF (or 3 × 47 pF);
4 × 1 µF; 1 × 4.7 µF; 1 × 47 µF;
2 × 3.3k; 1 × 10M; 1 × 100k;
1 × 1k; 2 × 10k

1 × 7476 JK flip-flop
1 × LF398 sample and hold
1 × 0.01 µF capacitor

1 × DAC0800 D to A converter
2 × 4.7k; 2x 10k
1 × 0.01 µF; 2x 0.1 µF

Part 3
2 × LM308N operational amplifier
1 × 741 operational amplifier
2 × 100 pF
2 × 2.2k; 1 × 56 Ω; 2 × 470 Ω;
2 × 4.7k

Printed in the United Kingdom
by Lightning Source UK Ltd.
103605UKS00001B/22